PROUST *and the* SQUID

PROUST *and* *the* SQUID

The STORY *and* SCIENCE *of the* READING BRAIN

MARYANNE WOLF

Illustrations by Catherine Stoodley

HARPER

An Imprint of HarperCollins*Publishers*
www.harpercollins.com

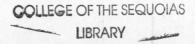

I dedicate this book to all the members
of my family . . . past, present, and still to come

HarperCollins books may be purchased for educational, business, or sales promotional use. For information, please write: Special Markets Department, HarperCollins Publishers, 10 East 53rd Street, New York, NY 10022.

FIRST EDITION

Designed by Renato Stanisic

Library of Congress Cataloging-in-Publication Data is available upon request.

ISBN: 978-0-06-018639-5
ISBN-10: 0-06-018639-9

07   08   09   10   11   DIX/RRD   10   9   8   7   6   5   4   3   2   1

CONTENTS

ILLUSTRATIONS

PREFACE

I HAVE LIVED MY LIFE IN THE SERVICE OF WORDS: finding where they hide in the convoluted recesses of the brain, studying their layers of meaning and form, and teaching their secrets to the young. In these pages I invite you to ponder the profoundly creative quality at the heart of reading words. Nothing in our intellectual development should be less taken for granted at this moment in history, as the transition to a digital culture accelerates its pace.

This is particularly so because there has also never been a time when the complex beauty of the reading process stood more revealed, when the magnitude of its contributions was more clearly understood by science, or when these contributions seemed more in danger of being replaced by new forms of communication. Examining what we have and reflecting on what we want to preserve are the leitmotifs of these pages.

To truly understand what we do when we read would be, as the fin de siècle scholar Sir Edmund Huey memorably wrote long ago, "the acme of a psychologist's achievements, for it would be to describe very many of the most intricate workings of the human mind, as well as to unravel the tangled story of the most remarkable specific performance that civilization has learned in all its history." Informed by areas of study as varied as evolutionary history and cognitive neuroscience, our contemporary knowledge about the reading brain would have dazzled Huey. We know that each new type of writing system was developed through millennia of human history, and required different adaptations of the human brain; we know that the multifaceted development of reading extends from infancy to ever-deepening levels of expertise; and we know that the curious mix of challenge and gift to be found in dyslexia—in which the brain struggles to learn to read—contains insights that are transforming our understanding of reading. Together, these areas of knowledge illuminate the brain's nearly miraculous capacity to rearrange itself to learn to read, and in the process to form new thoughts.

In this book I hope to push you gently toward reconsidering things you might long have taken for granted—such as how natural it is for a child to learn to read. In the evolution of our brain's capacity to learn, the act of reading is not natural, with consequences both marvelous and tragic for many people, particularly children.

To narrate this book demands a set of perspectives that have taken me many years to prepare for. I am a teacher of child development and cognitive neuroscience; a researcher of language, reading, and dyslexia; a parent of children you will learn about; and an apologist for written language. I direct a research center, the Center for Reading and Language Research, in the Eliot-Pearson Department of Child Development at Tufts University in Boston, where my colleagues and I conduct research on readers of all ages, particularly those with dyslexia. Together, we study what it means to be dyslexic in languages around the world, from languages that share roots with English—like German, Spanish, Greek, and Dutch—to less related languages like Hebrew, Japanese, and Chinese. We know the toll that not learning to read takes on children regardless of their native language, whether in struggling Filipino communities, on Native American reservations, or in affluent Boston suburbs. Many of our efforts explore the design of new interventions and the effects of these interventions on behaviors in the classroom and in the brain. Thanks to imaging technology, we can actually "see" how the brain reads before and after our work is done.

The sum of these experiences, the amount of research available, and the recognition of society's shift into new modes of communication compelled me to write my first book for the general public. I am, it must be said, still becoming accustomed to a style where there is no immediate reference to the many scholars whose research underlies so much of this book. I earnestly hope the reader will take advantage of the extensive notes and references that accompany each chapter.

The book begins by celebrating the beauty, variety, and transformative capacities of the origins of writing; proceeds to the dramatic new landscapes of the development of the reading brain and its various pathways to acquisition; and ends with difficult questions about the virtues and dangers in what lies ahead.

Oddly enough, a preface often presents the author's final thoughts to the reader on finishing the book. This book is no exception. But rather than end with my own words, I wish to use those from the gentle curator of Marilynne Robinson's *Gilead*, as he gave his best writings to his young son: "I wrote almost all of it in the deepest hope and conviction. Sifting my thoughts and choosing my words. Trying to say what was true. And I'll tell you frankly, that was wonderful."

PART I

HOW *the* BRAIN
LEARNED *to* READ

Words and music are the tracks of human evolution.
—JOHN S. DUNNE

*Knowing how something originated often is
the best clue to how it works.*
—TERRENCE DEACON

Chapter 1

READING LESSONS FROM PROUST AND THE SQUID

I believe that reading, in its original essence, [is] that fruitful miracle of a communication in the midst of solitude.
—MARCEL PROUST

Learning involves the nurturing of nature.
—JOSEPH LEDOUX

WE WERE NEVER BORN TO READ. HUMAN BEINGS invented reading only a few thousand years ago. And with this invention, we rearranged the very organization of our brain, which in turn expanded the ways we were able to think, which altered the intellectual evolution of our species. Reading is one of the single most remarkable inventions in history; the ability to record history is one of its consequences. Our ancestors' invention could come about only because of the human brain's extraordinary ability to make new connections among its existing structures, a process made possible by the brain's ability to be shaped by experience. This plasticity at the heart of the brain's design forms the basis for much of who we are, and who we might become.

This book tells the story of the reading brain, in the context of our unfolding intellectual evolution. That story is changing before our eyes and under the tips of our fingers. The next few decades will witness transformations in our ability to communicate, as we recruit new connections in the brain that will propel our intellectual development in new and different ways. Knowing what reading demands of our brain and knowing how it contributes to our capacity to think, to feel, to infer, and to understand other human beings is especially important today as we make the transition from a reading brain to an increasingly digital one. By coming to understand how reading evolved historically, how it is acquired by a child, and how it restructured its biological underpinnings in the brain, we can shed new light on our wondrous complexity as a literate species. This places in sharp relief what may happen next in the evolution of human intelligence, and the choices we might face in shaping that future.

This book consists of three areas of knowledge: the early history of how our species learned to read, from the time of the Sumerians to Socrates; the developmental life cycle of humans as they learn to read in ever more sophisticated ways over time; and the story and science of what happens when the brain can't learn to read. Taken together, this cumulative knowledge about reading both celebrates the vastness of our accomplishment as the species that reads, records, and goes beyond what went before, and directs our attention to what is important to preserve.

There is something less obvious that this historical and evolutionary view of the reading brain gives us. It provides a very old and very new approach to how we teach the most essential aspects of the reading process—both for those whose brains are poised to acquire it and for those whose brains have systems that may be organized differently, as in the reading disability known as dyslexia. Understanding these unique hardwired systems—which are preprogrammed generation after generation by instructions from our genes—advances our knowledge in unexpected ways that have implications we are only beginning to explore.

Interwoven through the book's three parts is a particular view of how the brain learns anything new. There are few more powerful mirrors of the human brain's astonishing ability to rearrange itself to learn a new intellectual function than the act of reading. Underlying the brain's ability to learn reading lies its protean capacity to make new connections among structures and circuits originally devoted to other more basic brain processes that have enjoyed a longer existence in human evolution, such as vision and spoken language. We now know that groups of neurons create new connections and pathways among themselves every time we acquire a new skill. Computer scientists use the term "open architecture" to describe a system that is versatile enough to change—or rearrange—to accommodate the varying demands on it. Within the constraints of our genetic legacy, our brain presents a beautiful example of open architecture. Thanks to this design, we come into the world programmed with the capacity to change what is given to us by nature, so that we can go beyond it. We are, it would seem from the start, genetically poised for breakthroughs.

Thus the reading brain is part of highly successful two-way dynamics. Reading can be learned only because of the brain's plastic design, and when reading takes place, that individual brain is forever changed, both physiologically and intellectually. For example, at the neuronal level, a person who learns to read in Chinese uses a very particular set of neuronal connections that differ in significant ways from the pathways used in reading English. When Chinese readers first try to read in English, their brains attempt to use Chinese-based neuronal pathways. The act of learning to read Chinese characters has literally shaped the Chinese reading brain. Similarly, much of how we think and what we think about is based on insights and associations generated from what we read. As the author Joseph Epstein put it, "A biography of any literary person ought to deal at length with what he read and when, for in some sense, *we are what we read*."

These two dimensions of the reading brain's development and evolution—the personal-intellectual and the biological—are rarely described together, but there are critical and wonderful lessons to be discovered in doing just that. In this book I use the celebrated French novelist Marcel Proust as metaphor and the largely underappreciated squid as analogy for two very different aspects of reading. Proust saw reading as a kind of intellectual "sanctuary," where human beings have access to thousands of different realities they might never encounter or understand otherwise. Each of these new realities is capable of transforming readers' intellectual lives without ever requiring them to leave the comfort of their armchairs.

Scientists in the 1950s used the long central axon of the shy but cunning squid to understand how neurons fire and transmit to each other, and in some cases to see how neurons repair and compensate when something goes awry. At a different level of study, cognitive neuroscientists today investigate how various cognitive (or mental) processes work in the brain. Within this research, the reading process offers an example par excellence of a recently acquired cultural invention that requires something new from existing structures in the brain. The study of what the human brain has to do to read, and of its clever ways of adapting when things go wrong, is analogous to the study of the squid in earlier neuroscience.

Proust's sanctuary and the scientist's squid represent complementary ways of understanding different dimensions in the reading process. Let me introduce you more concretely to the approach of this book by having you read two of Proust's breath-defying sentences from his book *On Reading*, as fast as you can.

> There are perhaps no days of our childhood we lived
> so fully as those . . . we spent with a favorite book.
> Everything that filled them for others, so it seemed, and
> that we dismissed as a vulgar obstacle to a *divine plea-*
> *sure*: the game for which a friend would come to fetch us
> at the most interesting passage; the troublesome bee or
> sun ray that forced us to lift our eyes from the page or to
> change position; the provisions for the afternoon snack
> that we had been made to take along and that we left
> beside us on the bench without touching, while above our
> head the sun was diminishing in force in the blue sky; the
> dinner we had to return home for, and during which we
> thought only of going up immediately afterward to finish

the interrupted chapter, all those things with which reading should have kept us from feeling anything but annoyance, on the contrary they have engraved in us so sweet a memory (so much more precious to our present judgment than what we read then with such love), that if we still happen today to leaf through those books of another time, it is for no other reason than that they are the only calendars we have kept of days that have vanished, and we hope to see reflected on their pages the dwellings and the ponds which no longer exist.

Consider first what you were thinking while reading this passage, and then try to analyze exactly what you did as you read it, including how you began to connect Proust to other thoughts. If you are like me, Proust conjured up your own long-stored memories of books: the secret places you found to read away from the intrusions of siblings and friends; the thrilling sensations elicited by Jane Austen, Charlotte Brontë, and Mark Twain; the muffled beam of the flashlight you hoped your parent wouldn't notice beneath the sheets. This is Proust's reading sanctuary, and it is ours. It is where we first learned to roam without abandon through Middle Earth, Lilliput, and Narnia. It is the place we first tried on the experiences of those we would never meet: princes and paupers, dragons and damsels, !Kung warriors, and a Dutch-Jewish girl hiding with her family from Nazi soldiers.

It is said that Machiavelli would sometimes prepare to read by dressing up in the period of the writer he was reading and then setting a table for the two of them. This was his sign of respect for the author's gift, and perhaps of Machiavelli's tacit understanding of the sense of encounter that Proust described. While reading, we can leave our own consciousness, and pass over into the consciousness of another person, another age, another culture. "Passing over," a term used by the theologian John Dunne, describes the process through which reading enables us to try on, identify with, and ultimately enter for a brief time the wholly different perspective of another person's consciousness. When we pass over into how a knight thinks, how a slave feels, how a heroine behaves, and how an evildoer can regret or deny wrongdoing, we never come back quite the same; sometimes we're inspired, sometimes saddened, but we are always enriched. Through this exposure we learn both the commonality and the uniqueness of our own thoughts—that we are individuals, but not alone.

The moment this happens, we are no longer limited by the confines of our own thinking. Wherever they were set, our original boundaries are challenged, teased, and gradually placed somewhere new. An expanding sense of "other" changes who we are, and, most importantly for children, what we imagine we can be.

Let's go back to what you did when I asked you to switch your attention from this book to Proust's passage and to read as fast as you could without losing Proust's meaning. In response to this request, you engaged an array of mental or cognitive processes: attention; memory; and visual, auditory, and linguistic processes. Promptly, your brain's attentional and executive systems began to plan how to read Proust speedily and still understand it. Next, your visual system raced into action, swooping quickly across the page, forwarding its gleanings about letter shapes, word forms, and common phrases to linguistic systems awaiting the information. These systems rapidly connected subtly differentiated visual symbols with essential information about the sounds contained in words. Without a single moment of conscious awareness, you applied highly automatic rules about the sounds of letters in the English writing system, and used a great many linguistic processes to do so. This is the essence of what is called the alphabetic principle, and it depends on your brain's uncanny ability to learn to connect and integrate at rapid-fire speeds what it sees and what it hears to what it knows.

As you applied all these rules to the print before you, you activated a battery of relevant language and comprehension processes with a rapidity that still astounds researchers. To take one example from the language domain, when you read the 233 words in Proust's passage, your word meaning, or semantic, systems contributed every possible meaning of each word you read and incorporated the exact correct meaning for each word in its context. This is a far more complex and intriguing process than one might think. Years ago, the cognitive scientist David Swinney helped uncover the fact that when we read a simple word like "bug," we activate not only the more common meaning (a crawling, six-legged creature), but also the bug's less frequent associations—spies, Volkswagens, and glitches in software. Swinney discovered that the brain doesn't find just one simple meaning for a word; instead it stimulates a veritable trove of knowledge about that word and the many words related to it. The richness of this semantic dimension of reading depends on the riches we have already stored, a fact with important and sometimes devastating developmental implications for our children. Children with a rich repertoire of words and their associations will experience any text or any conversation in ways that are substantively different from children who do not have the same stored words and concepts.

Think about the implications of Swinney's finding for texts as simple as Dr. Seuss's *Oh, The Places You'll Go!* or as semantically complex as James Joyce's *Ulysses*. Children who have never left the narrow boundaries of their neighborhood, either figuratively or literally, may understand this book in entirely different ways from other children. We bring our entire store of meanings to whatever we read—or not. If we apply this finding to the passage from Proust that you just read, it means that your executive planning system directed a great many activities to ensure that you comprehended what was there, and retrieved all your personal associations to the text. Your grammatical system had to work overtime to avoid stumbling over Proust's unfamiliar sentence constructions, like his use of long clauses strung together by many commas and semicolons before the predicate. To accomplish all this without forgetting what you already read fifty words back, your semantic and grammatical systems had to function closely with your working memory. (Think of this type of memory as a kind of "cognitive blackboard," which temporarily stores information for you to use in the near term.) Proust's unusually sequenced grammatical information had to be connected to the meanings of individual words without losing track of the overall propositions and context of the passage.

As you linked all this linguistic and conceptual information, you generated your own inferences and hypotheses based on your own background knowledge and engagement. If this cumulative information failed to make sense, you might have reread some parts to ensure that they fit within the given context. Then, after you integrated all this visual, conceptual, and linguistic information with your background knowledge and inferences, you arrived at an understanding of what Proust was describing: a glorious day in childhood made timeless through the "divine pleasure" that is reading!

Then, some of you paused at the end of Proust's passage and went somewhere beyond what the text provided. But before tackling this more philosophical point, let's turn back to the biological dimension and look immediately below the surface of the behavioral act of reading. All human behaviors rest on layers on layers of teeming, underlying activity. I asked the neuroscientist and artist Catherine Stoodley of Oxford to draw a pyramid to illustrate how these various levels operate together when we read a single word (Figure 1-1). In the top layer of this pyramid, reading the word "bear" is the surface behavior; below it is the cognitive level, which consists of all those basic attentional, perceptual, conceptual, linguistic, and motor processes you just used to read. These cognitive processes, which many psychologists spend their entire lives studying, rest on tangible neurological structures that are made up of neurons built up and then guided by the interaction between genes and the environment. In other words, all human *behaviors* are based on multiple *cognitive* processes, which are based on the rapid integration of information from very specific *neurological structures*, which rely on billions of *neurons* capable of trillions of possible connections, which are programmed in large part by *genes*. In order to learn to work together to perform our most basic human functions, neurons need instructions from genes about how to form efficient *circuits* or *pathways* among the neurological structures.

This pyramid functions like a three-dimensional map for understanding how any genetically programmed behavior, such as vision, happens. It does not explain, however, how it can be ap-

Figure 1-1: Pyramid of Reading

plied to a reading circuit, because there are no genes specific only to reading in the bottom layer. Unlike its component parts such as vision and speech, which *are* genetically organized, reading has no direct genetic program passing it on to future generations. Thus the next four layers involved must learn how to form the necessary pathways anew every time reading is acquired by an individual brain. This is part of what makes reading—and any cultural invention—different from other processes, and why it does not come as naturally to our children as vision or spoken language, which are preprogrammed.

How, then, did the first time ever occur? The French neuroscientist Stanislas Dehaene tells us that the first humans who invented writing and numeracy were able to do this by what he calls "neuronal recycling." For example, in his work with primates,

Dehaene shows that if you put two plates of bananas in front of a monkey—one with two bananas and one with four—an area in the monkey's posterior cortex will activate just before he grasps the more bountiful plate. This same general area is one of the regions of the brain we humans now use for some mathematical operations. Similarly, Dehaene and his colleagues argue that our ability to recognize words in reading uses the species' evolutionarily older circuitry that is specialized for object recognition. Furthermore, just as our ancestors' capacity to distinguish between predator and prey at a glance drew on an innate capacity for visual specialization, our ability to recognize letters and words may involve an even further in-built capacity that allows "specialization within a specialization."

If one were to expand Dehaene's view somewhat, it would seem more than likely that the reading brain exploited older neuronal pathways originally designed not only for vision but for connecting vision to conceptual and linguistic functions: for example, connecting the quick recognition of a shape with the rapid inference that this footprint can signal danger; connecting a recognized tool, predator, or enemy with the retrieval of a word. When confronted, therefore, with the task of inventing functions like literacy and numeracy, our brain had at its disposal three ingenious design principles: the capacity to make new connections among older structures; the capacity to form areas of exquisitely precise specialization for recognizing patterns in information; and the ability to learn to recruit and connect information from these areas automatically. In one way or another, these three principles of brain organization are the foundation for all of reading's evolution, development, and failure.

The elegant properties of the visual system provide an excellent example of how recycling existing visual circuits made the development of reading possible. Visual cells possess the capacity to become highly specialized and highly specific, and to make new circuits among preexisting structures. This allows babies to come into the world with eyes that are almost ready to fire and that are exceptional examples of design and precision. Soon after birth, each neuron in the eye's retina begins to correspond to a specific set of cells in the occipital lobes. Because of this design feature in our visual system, called retinotopic organization, every line, diagonal, circle, or arc seen by the retina in the eye activates a specific, specialized location in the occipital lobes in a split second (Figure 1-2).

This quality of the visual system is somewhat different from why our Cro-Magnon ancestors could identify animals on the distant horizon, why many of us can identify the model of a car a quarter-mile away, and why bird-watchers can identify a tern other people may not even see. Dehaene suggests that the visual areas in our ancestors' brains responsible for object recognition

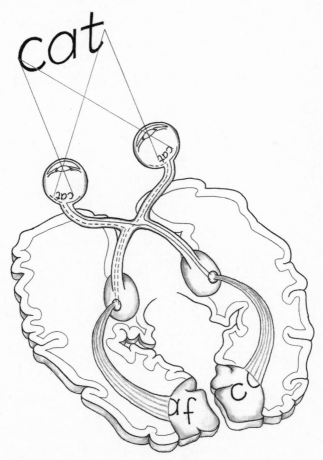

Figure 1-2: Visual Systems

were used to decipher the first symbols and letters of written language by adapting their built-in system for recognition. Critically, the combination of several innate capacities—for adaptation, for specialization, and for making new connections—allowed our brain to make new pathways between visual areas and those areas serving the cognitive and linguistic processes that are essential to written language.

The third principle exploited by reading—the capacity of the neuronal circuits to become virtually automatic—incorporates the other two. This is what allowed you to fly across Proust's passage and understand what you read. Becoming virtually automatic does not happen overnight and is not a characteristic of either a novice bird-watcher or a young novice reader. These circuits and pathways are created through hundreds or, in the case of some children with reading disabilities like dyslexia, thousands of exposures to letters and words. The neuronal pathways for recognizing letters, letter patterns, and words become automatic thanks to retinotopic organization, object recognition capacities, and to one other extremely important dimension of brain organization: our ability to *represent* highly learned patterns of information in our specialized regions. For example, as the networks of cells responsible for recognizing letters and letter patterns learn to "fire together," they create representations of their visual information that are far more rapidly retrieved.

Fascinatingly, networks of cells that have learned to work together over a long time produce representations of visual information, even when this information isn't in front of us. In an illuminating experiment by Harvard cognitive scientist Stephen Kosslyn, adult readers in a brain scanner were asked to close their eyes and imagine certain letters. When they were asked about capital letters, discrete regions responsible for one part of the visual field in the visual cortex responded; lowercase letters triggered other discrete areas. Thus merely imagining letters results in activation of particular neurons in our visual cortex. For the expert reading brain, as information enters through the retina, all the physical properties of the letters are processed by an array of specialized neurons that feed their information automatically deeper and deeper into other visual processing areas. They are part and parcel of the virtual automaticity of the reading brain, in which all its representations and indeed all its individual processes—not just visual ones—become rapid-fire and effortless.

What happens between our first exposure to letters and expert reading is very important to scientists because it offers a unique opportunity to watch the orderly development of a cognitive process. The various features that characterize the visual system— enlisting older genetically programmed structures, recognizing patterns, creating discrete working groups of specialized neurons for particular representations, making circuit connections with great versatility, and achieving fluency through practice—are similar in all the other major cognitive and linguistic systems involved in reading. I will elaborate on this later, but first I want to highlight a marvelous (and hardly coincidental) analogue between what happens in the brain and what happens in the internal thoughts of every reader.

In much the way reading reflects the brain's capacity for going beyond the original design of its structures, it also reflects the reader's capacity to go beyond what is given by the text and the author. As your brain's systems integrated all the visual, auditory, semantic, syntactic, and inferential information from Proust's passage about a single day in childhood with a beloved book, you, the reader, automatically began to connect what Proust wrote with your own thinking and personal insights.

I cannot, of course, describe where your thoughts went, but I can describe mine. Because I had just visited an exhibit at the Boston Museum of Fine Arts on Monet and impressionism, I found myself connecting how Proust wrote about a single day in his childhood with how Monet painted *Impression: Sunrise*. Both Proust and Monet used pieces of information to render a composite that made a more vivid impression than if they had created a perfect reproduction. In so doing, both artist and novelist are examples of Emily Dickinson's enigmatic charge to "tell all the Truth, but tell it slant—/Success in Circuit lies."

Emily Dickinson never envisioned neuronal circuits when she wrote those lines, but it turns out that she was as astute physiologically as she was poetically. By using indirect approaches, Proust and Monet force their readers and viewers to contribute actively to the constructions themselves, and in the process to experience them more directly. Reading is a neuronally and intellectually circuitous act, enriched as much by the unpredictable indirections of a reader's inferences and thoughts, as by the direct message to the eye from the text.

This unique aspect of reading has begun to trouble me considerably as I consider the Google universe of my children. Will the constructive component at the heart of reading begin to change and potentially atrophy as we shift to computer-presented text, in which massive amounts of information appear instantaneously? In other words, when seemingly complete visual information is given almost simultaneously, as it is in many digital presentations, is there either sufficient time or sufficient motivation to process the information more inferentially, analytically, and critically? Is the act of reading dramatically different in such contexts? The basic visual and linguistic processes might be identical, but would the more time-demanding, probative, analytical, and creative aspects of comprehension be foreshortened? Or does the potential added information from hyperlinked text contribute to the development of children's thinking? Can we preserve the constructive dimension of reading in our children alongside their growing abilities to perform multiple tasks and to integrate ever-expanding amounts of information? Should we begin to provide explicit instruction for reading multiple modalities of text presentation to ensure that our children learn multiple ways of processing information?

I stray with these questions. But indeed we stray often when we read. Far from being negative, this associative dimension is part of the generative quality at the heart of reading. One hundred fifty years ago Charles Darwin saw in creation a similar principle, whereby "endless" forms evolve from finite principles: "From so simple a beginning, endless forms most beautiful and most wonderful have been, and are being evolved." So it is with written language. Biologically and intellectually, reading allows the species to go "beyond the information given" to create endless thoughts most beautiful and wonderful. We must not lose this essential quality in our present moment of historical transition to new ways of acquiring, processing, and comprehending information.

To be sure, the relationship between readers and text differs across cultures and across history. Thousands of lives have been altered or lost depending on whether a sacred text like the Bible is read in a concrete, literal way or in a generative, interpretative way. Martin Luther's act of translating the Latin Bible into the German language, which permitted ordinary people to read and interpret it for themselves, significantly influenced the history of religion. Indeed, as some historians observe, the changing relationship of readers to text over time can be seen as one index of the history of thought.

The thrust of this book, however, will be more biological and cognitive than cultural-historical. Within that context, the generative capacity of reading parallels the fundamental plasticity in the circuit wiring of our brains: both permit us to go beyond the particulars of the given. The rich associations, inferences, and insights emerging from this capacity allow, and indeed invite, us to reach beyond the specific content of what we read to form new thoughts. In this sense reading both reflects and reenacts the brain's capacity for cognitive breakthroughs.

Proust said most of this, if more obliquely, in a powerful description of the ability of reading to elicit our own thinking.

> We feel quite truly that our wisdom begins where that of the author ends, and we would like to have him give us answers, while all he can do is give us desires. And these desires he can arouse in us only by making us contemplate the supreme beauty which the last effort of his art has permitted him to reach. But by . . . a law which perhaps signifies that we can receive the truth from nobody, and that we must create it ourselves, that which is the end of their wisdom appears to us as but the beginning of ours.

Proust's understanding of the generative nature of reading contains a paradox: the goal of reading is to go beyond the author's ideas to thoughts that are increasingly autonomous, transformative, and ultimately independent of the written text. From the child's first, halting attempts to decipher letters, the experience of reading is not so much an end in itself as it is our best vehicle to a transformed mind, and, literally and figuratively, to a changed brain.

Ultimately, the biological and intellectual transformations brought about by reading provide a remarkable petri dish for examining how we think. Such an examination requires multiple perspectives—ancient and modern linguistics, archaeology, history, literature, education, psychology, and neuroscience. The goal of this book is to integrate these disciplines to present new perspectives on three aspects of written language: the evolution of the reading brain (how the human brain learned to read); its development (how the young brain learns to read and how reading changes us); and its variations (when the brain can't learn to read).

How the Brain Learned to Read

We will begin in Sumer, Egypt, and Crete, where the still mysterious beginnings of written language can be found among Sumerian cuneiform, Egyptian hieroglyphs, and some recently discovered proto-alphabetic scripts. Each major type of writing invented by our ancestors demanded something a little different from the brain, and this may explain why more than 2,000 years elapsed between these earliest known writing systems and the remarkable, almost perfect alphabet developed by the ancient Greeks. At its root the alphabetic principle represents the profound insight that each word in spoken language consists of a finite group of individual sounds that can be represented by a finite group of individual letters. This seemingly innocent-sounding principle was totally revolutionary when it emerged over time, for it created the capacity for every spoken word in every language to be translated into writing.

Why Socrates directed all his legendary rhetorical skills against the Greek alphabet and the acquisition of literacy is one of the great, largely untold stories in the history of reading. In words unerringly prescient today, Socrates described what would be lost to human beings in the transition from oral to written culture. Socrates' protests—and the silent rebellion of Plato as he recorded every word—are notably relevant today as we and our children negotiate our own transition from a written culture to one that is increasingly driven by visual images and massive streams of digital information.

How the Young Brain Learns to Read and How We Are Changed over the Life Span

Several thought-provoking links connect the history of writing in the species to the development of reading in the child. The first is the fact that although it took our species roughly 2,000 years to make the cognitive breakthroughs necessary to learn to read with an alphabet, today our children have to reach those same insights about print in roughly 2,000 days. The second concerns the evolutionary and educational implications of having a "re-arranged" brain for learning to read. If there are no genes specific only to reading, and if our brain has to connect older structures for vision and language to learn this new skill, every child in every generation has to do a lot of work. As the cognitive scientist Steven Pinker eloquently remarked, "Children are wired for sound, but print is an optional accessory that must be painstakingly bolted on." To acquire this unnatural process, children need instructional environments that support all the circuit parts that need bolting for the brain to read. Such a perspective departs from current teaching methods that focus largely on only one or two major components of reading.

Understanding the period in development stretching from infancy to young adulthood necessitates an understanding of the full range of circuit parts in the reading brain and their development. It also involves a tale of two children, both of whom must acquire hundreds upon hundreds of words, thousands of concepts, and tens of thousands of auditory and visual perceptions. These are the raw materials for developing the major components of reading. Owing largely to their environments, however, one child will acquire these essentials, and the other will not. Through no fault of their own, the needs of thousands of children go unmet every day.

Learning to read begins the first time an infant is held and read a story. How often this happens, or fails to happen, in the first five years of childhood turns out to be one of the best predictors of later reading. A little-discussed class system invisibly divides our society, with those families that provide their children environments rich in oral and written language opportunities gradually set apart from those who do not, or cannot. A prominent study found that by kindergarten, a gap of 32 million words already separates some children in linguistically impoverished homes from their more stimulated peers. In other words, in some environments the average young middle-class child hears 32 million more spoken words than the young underprivileged child by age five.

Children who begin kindergarten having heard and used thousands of words, whose meanings are already understood, classified, and stored away in their young brains, have the advantage on the playing field of education. Children who never have a story read to them, who never hear words that rhyme, who never imagine fighting with dragons or marrying a prince, have the odds overwhelmingly against them.

Knowledge about the precursors of reading can help change that situation. Thanks to remarkable new technologies, we can now see what happens if all goes right in the acquisition of reading, as a child moves from decoding a word like "cat" to the fluent, seemingly effortless comprehension of "a feline creature named Mephistopheles." We find a series of predictable phases that a human passes through across the life span, illustrating just how different the circuits and requirements of a new reader's brain are from those of an expert reader, who navigates the tangled worlds of *Moby-Dick, War and Peace,* and texts on economics. Our growing knowledge about how the brain learns to read over time can help predict, ameliorate, and prevent some forms of unnecessary reading failure. Today, we possess sufficient knowledge about the components of reading to be able not only to diagnose almost every child in kindergarten at risk of a learning difficulty, but also to teach most children to read. This same knowledge underscores what we do not wish to lose in the achievement of the reading brain, just as the digital epoch begins to make new and different demands on that brain.

When the Brain Can't Learn to Read

Knowledge about reading failure provides a different angle on this knowledge base, with some surprises for anyone who looks there. From the viewpoint of science, dyslexia is a bit like studying a young squid that can't swim very fast. This squid's different wiring can teach us both about what is necessary for swimming and about the unique gifts this squid must have to be able to survive and flourish without swimming like every other squid. My colleagues and I use a variety of tools, from naming letters to brain imaging, to understand why so many children with dyslexia, including my own firstborn son, have difficulty not only with reading but also with seemingly simple linguistic behaviors like discriminating individual sounds or phonemes within words, or quickly retrieving the name of a color. By tracking activity in the brain as it performs these various behaviors in normal development and in dyslexia, we are constructing living maps of the neuronal landscape.

The surprises on this landscape increase daily. Recent advances in neuroimaging research begin to paint a different picture of the brain of a person with dyslexia that may have enormous implications for future research, and particularly for intervention. Understanding these advances can make the difference between having a huge number of our future citizens poised to contribute to society and having a huge number who cannot contribute what they could otherwise. Connecting what we know about the typical child's development to what we know about impediments in reading can help us reclaim the lost potential of millions of children, many of whom have strengths that could light up our lives.

For we are also in the exciting early stages of understanding the little-studied benefits that accompany the brain development of some persons with dyslexia. It is no longer reducible to coincidence that so many inventors, artists, architects, computer designers, radiologists, and financiers have a childhood history of dyslexia. The inventors Thomas Edison and Alexander Graham Bell, the business entrepreneurs Charles Schwab and David Neeleman, the artists Leonardo da Vinci and Auguste Rodin, and the Nobel prize–winning scientist Baruj Benacerraf are all extraordinarily successful individuals with a history of dyslexia or related reading disorders. What is it about the dyslexic brain that seems linked in some people to unparalleled creativity in their professions, which often involve design, spatial skills, and the recognition of patterns? Was the differently organized brain of a person with dyslexia better suited for the demands of the preliterate past, with its emphasis on building and exploring? Will individuals with dyslexia be even better suited to the visual, technology-dominated future? Is the most current imaging and genetic research giving us the outlines of a very unusual brain organization in some persons with dyslexia that may ultimately explain both their known weaknesses and our steadily growing understanding of their strengths?

Questions about the brain of a person with dyslexia lead us to look both backward to our evolutionary past and forward to the future of our symbolic development. What is being lost and what is being gained for so many young people who have largely replaced books with the multidimensioned "continuous partial attention" culture of the Internet? What are the implications of seemingly limitless information for the evolution of the reading brain and for us as a species? Does the rapid, almost instantaneous presentation of expansive information threaten the more time-demanding formation of in-depth knowledge? Recently, Edward Tenner, who writes about technology, asked whether Google promotes a form of information illiteracy and whether there may be unintended negative consequences of such a mode of learning: "It would be a shame if brilliant technology were to end up threatening the kind of intellect that produced it."

Reflecting on such questions underscores the value of intellectual skills facilitated through literacy that we don't wish to lose, just when we appear potentially poised to replace them with other skills. This book is two parts science, one part personal observation, and as much truth as I can find to tell about how fiercely we must work as a society to preserve the development of particular aspects of reading, both for this generation and for generations to come. I will argue that unlike Plato, who with deep ambivalence straddled oral language and literacy, we do not need to choose between two modes of communication; rather, we must be vigilant not to lose the profound generativity of the reading brain, as we add new dimensions to our intellectual repertoire.

Like Proust, however, I can lead the viewer only so far in the realm of established or given knowledge. My final chapter goes beyond the information that we know, into areas where we have only intuition and extrapolation to guide us. By the end of this exploration of the reading brain, what we know of the profound cognitive miracle that takes place every time a human being learns to read will be the reader's to preserve and to go beyond.

Chapter 2

HOW THE BRAIN ADAPTED ITSELF TO READ: THE FIRST WRITING SYSTEMS

And so I ambitiously proceed from my history as a reader to the history of the act of reading. Or rather, to a history of reading, since any such history—made up of particular institutions and private circumstances—must be only one of many.
—ALBERTO MANGUEL

The invention of writing, which occurred independently in distant parts of the world at many times, even occasionally in the modern era, must rank among mankind's highest intellectual achievements. Without writing, human culture as we know it today is inconceivable.
—O. TZENG AND W. WANG

LITTLE TOKENS IN HARDENED CLAY ENVELOPES, intricate dyed knots of twine in Incan *quipus* (see Figure 2-1), graceful designs scratched on the surface of turtle shells: the origins of writing took wondrously various shapes and forms over the last 10,000 years, all over the earth. Crosshatched lines on stones thought to be 77,000 years old were found recently under layers of earth in the Blomos Cave in South Africa and may prove to be still earlier signs of the first human efforts to "read."

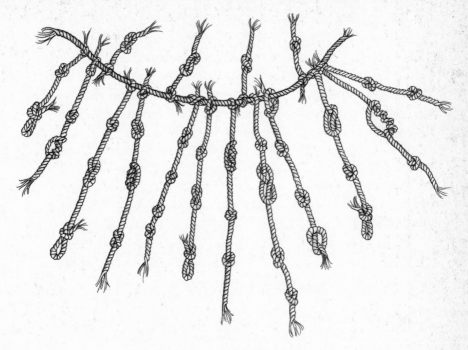

Figure 2-1: Example of Incan Quipus

Wherever and however it occurred, reading never "just happened." The story of reading reflects the sum of a series of cognitive and linguistic breakthroughs occurring alongside powerful cultural changes. Its colorful, spasmodic history helps reveal what our brain had to learn, one new process and insight at a time. It is a history not only of how we learned to read, but also of how different forms of writing required different adaptations of the brain's original structures and in the process helped to change the way we think. From the contemporary perspective of our own unfolding changes in communication, the story of reading offers a unique documentation of how each new writing system contributed something special to our species' intellectual development.

Across every known system, writing began with a set of two or more epiphanies. First came a new form of *symbolic representation*, one level of abstraction more than earlier drawings: the amazing discovery that simple marked lines on clay tokens, stones, or turtle shells can represent either something concrete in the natural world, such as a sheep; or something abstract, such as a number or an answer from an oracle. With the second breakthrough came the insight that a system of symbols can be used to communicate across time and space, preserving the words and thoughts of an individual or an entire culture. The third epiphany, the most linguistically abstract, did not happen everywhere: *sound-symbol correspondence* represents the stunning realization that all words are actually composed of tiny individual sounds and that symbols can physically signify each of these sounds for every word. Examining how several of our ancestors made these leaps into early writing provides us with a special lens on ourselves. Understanding the origins of a new process helps us see, as the neuroscientist Terry Deacon put it, "how it works." Understanding how it works, in turn, helps us know what we possess and what we need to preserve.

First, a Word about "Firsts"

No fewer than three recorded examples can be found of efforts by several monarchs to discover which was the first language spoken on earth. Herodotus tells us that the Egyptian king Psamtik I (664–610 BCE) ordered two newborn infants to be isolated in a shepherd's hut, with no exposure to human beings other than the shepherd, who brought in milk and food daily, or to any human language. Psamtik believed that the first words uttered by these infants would be the first language of the human race—a clever supposition, albeit false. Eventually, one infant cried out *bekos*, the Phrygian word for "bread." This single utterance inspired the long-held belief that Phrygian, spoken in northwest Anatolia, was the *Ursprache*, or original language, of humankind.

Centuries later, James IV of Scotland conducted a similar experiment, but with different, very interesting results: the Scottish infants involved "spake very guid Ebrew." On the European continent, Frederick II Hohenstaufen initiated another—and unfortunately more rigorous—experiment with two more infants, but both children died without speaking.

We may never say with authority which oral language came first, and there are even more questions about which written language was first. We can more easily, however, answer questions about whether writing was invented only once, or several times. In this chapter we trace several selected writing systems and how human beings learned to read everything from tiny tokens to "dragon bones" in the period that stretches from the eighth millennium to the first millennium BCE. Underlying this curious history is a less visible story of cerebral adaptation and change. With each of the new writing systems, with their different and increasingly sophisticated demands, the brain's circuitry rearranged itself, causing our repertoire of intellectual capacities to grow and change in great, wonderful leaps of thought.

The First Written Eureka: Symbolic Representation

By the mere fact of looking at these tablets, we have prolonged a memory from the beginning of our time, preserved a thought long after the thinker has stopped thinking, and made ourselves participants in an act of creation that remains open for as long as the incised images are seen, deciphered, read.
— ALBERTO MANGUEL

The chance discovery of little clay pieces, no larger than a quarter, marks the birth of modern efforts to learn about the history of writing. Called tokens, some of these pieces came enclosed in clay envelopes (see Figure 2-2) that bore markings representing their contents. We now know that these pieces date back to the period between 8000 and 4000 BCE, and formed a kind of accounting system used across many parts of the ancient world. The tokens primarily recorded the number of goods bought or sold, such as sheep, goats, and bottles of wine. A lovely irony of our species' cognitive growth is that the world of letters may have begun as an envelope for the world of numbers.

Simultaneously, the development of numbers and letters pro-

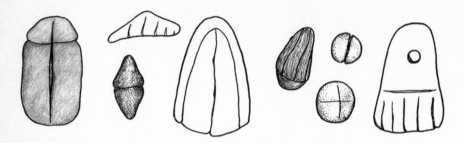

Figure 2-2: Tokens

moted both ancient economies and our ancestors' intellectual skills. For the first time "stock" could be counted with no necessity for the sheep, goats, or wine to be present. The precursor of stored data, a permanent record came into existence, accompanied by new cognitive capacities. For example, along with cave drawings like those in France and Spain, tokens reflected the emergence of a new human ability: the use of a form of symbolic representation, in which objects could be symbolized by marks for the eye. To "read" a symbol demanded two sets of novel connections: one cognitive-linguistic and the other cerebral. Among the long-established brain circuits for vision, language, and conceptualization, new connections developed and new retinotopic pathways—between the eye and specialized visual areas—became assigned to the tiny token marks.

We will never have a brain scan of our ancestors reading a token, but by using present knowledge of brain function, we can extrapolate to make a decent guess about what went on in their brains. The neuroscientists Michael Posner and Marcus Raichle, and Raichle's research group at Washington University, conducted a pioneering series of brain imaging studies to observe what the brain does when we look at a continuum of symbol-like characters with and without meaning. Their range of tasks included meaningless symbols, meaningful symbols that make up real letters, meaningless words, and meaningful words. Although clearly designed for other purposes, these studies provide a remarkable glimpse of what happens when the brain encounters ever more abstract and demanding writing systems—millennia ago and today.

Raichle's group found that when humans look at lines which convey no meaning, we activate only limited visual areas located in the occipital lobes at the back of the brain. This finding exemplifies some aspects of retinotopic organization, mentioned in Chapter 1. The cells in the retina activate a group of specific cells in the occipital regions that correspond to discrete visual features such as lines and circles.

When we see these same circles and lines and interpret them as meaningful symbols, however, we need new pathways. As Raichle's work showed, the presence of real-word status and meaning doubles or triples the brain's neuronal activity. Becoming familiar with the basic pathways used in a token-reading brain is an excellent foundation for understanding what happens in more complex reading brains. Our ancestors could read tokens because their brains were able to connect their basic visual regions to adjacent regions dedicated to more sophisticated visual and conceptual processing. These adjacent regions are found in other occipital and nearby temporal and parietal areas of the brain. The temporal lobes are involved in an impressive range of auditory and language-based processes, which contribute to our ability to comprehend the meanings of words. Similarly, the parietal lobes participate in a wide range of language-related processes, as well as spatial and computational functions. When a visual symbol like a token is imbued with meaning, our brain connects the basic visual areas to both the language system and the conceptual system in the temporal and parietal lobes and also to visual and auditory specialization regions called "association areas."

Symbolization, therefore, even for the tiny token, exploits and expands two of the most important features of the human brain—our capacity for specialization and our capacity for making new connections among association areas. One major difference between the human brain and the brain of any other primate is the proportion of our brain devoted to these association areas. Essential to reading symbols, these areas are responsible both for more demanding sensory processing and for making mental representations of information for future use (think of "representations"). Such a representational capacity is profoundly

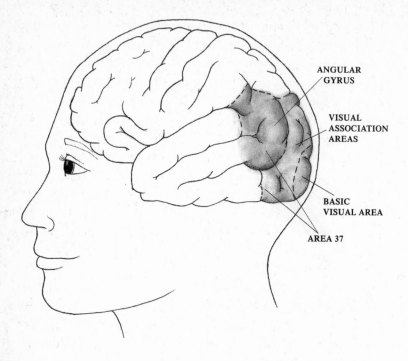

ANGULAR
GYRUS

VISUAL
ASSOCIATION
AREAS

BASIC
VISUAL AREA

AREA 37

Figure 2-3: First "Token-Reading Brain"

important for the use of symbols and for much of our intellectual life. It helps humans remember and retrieve stored representations of all sorts, from visual images like predators' footprints and tokens to auditory sounds like words and a tiger's growl. Further, this representational ability prepares the foundation for our evolutionary capacity to become virtually automatic at recognizing patterns in information all about us. All this enables us to become specialists in identifying various sensory information—whether tracks of woolly mammoths or tokens for goats. It is all of a piece.

Reading symbols required more than the visual specialization of our ancestors. Linking visual representations to linguistic and conceptual information was critical. Located at the juncture of the three posterior lobes of the brain, the *angular gyrus* area, described as the "association area of the association areas" by the great behavioral neurologist Norman Geschwind, is ideally located for linking different kinds of sensory information. The nineteenth-century French neurologist Joseph-Jules Déjerine observed that an injury to the angular gyrus region produced a loss in reading and writing. And today, the neuroscientists John Gabrieli and Russ Poldrack and their groups at MIT and UCLA find, through their imaging research, that pathways to and from the angular gyrus region become intensely activated during reading development.

From Raichle's, Poldrack's, and Gabrieli's work, we can infer that the likely physiological basis of our ancestors' first reading of tokens was a tiny new circuitry connecting the angular gyrus region with a few nearby visual areas and if Dehaene is correct, a few parietal areas involved in numeracy and occipital-temporal areas involved in object recognition (i.e., area 37) (Figure 2-3). However rudimentary, a novel form of connectivity began with the use of tokens, and with it came our species' earliest cognitive breakthrough in reading. By teaching new generations to use an increasing repertoire of symbols, our ancestors essentially passed on knowledge about the brain's capability for adaptation and change. Our brain was preparing to read.

From the Mouths of Kings and Queens: The Second Breakthrough into Cuneiform and Hieroglyphic Writing Systems

Have you noticed how picturesque the letter Y is and how innumerable its meanings are? The tree is a Y, the junction of two roads forms a Y, two converging rivers, a donkey's head and that of an ox, the glass with its stem, the lily on its stalk and the beggar lifting his arms are a Y. This observation can be extended to everything that constitutes the elements of the various letters devised by man.
— VICTOR HUGO

Toward the end of the fourth millennium BCE (3300–3200), a second breakthrough occurred: individual Sumerian inscriptions developed into a cuneiform system and Egyptian symbols became a

Figure 2-4: Examples of Cuneiform

hieroglyphic system. Whether or not the Sumerians or the Egyptians are the inventors of writing is increasingly debated. But there is no debate that the Sumerians invented one of the first and most revered systems for writing, whose influence continued through the great Akkadian system throughout all of Mesopotamia. The word "cuneiform" derives from the Latin word *cuneus*, "nail," which refers to the script's wedge-like appearance. Using a pointed reed stylus on soft clay, our ancestors created a script that looks, to the untutored eye, a lot like bird tracks (Figure 2-4).

The discovery of these strange-looking symbols is comparatively recent and a testimony to the lengths to which some intrepid linguists go to understand the origins of language. Contemporary language scholars like to recount how the nineteenth-century scholar and soldier Henry Raulinson risked life and limb to examine ancient writing in what is now Iran. Raulinson dangled on a rope 300 feet in the air to copy some of the first Sumerian writings carved into the side of a cliff.

Mercifully, the 5,000 extant Sumerian tablets are much more

accessible. Found in Sumerian palaces, temples, and warehouses, this writing was originally invented and used largely for administration and accounting. The ancient inhabitants of the Tigris-Euphrates delta themselves had a far more romantic notion of how their writing came to be. In one epic tale a messenger from the lord of Kulab arrived at a distant kingdom, too exhausted to deliver an important oral message. So as not to be frustrated by mortal failings, the lord of Kulab had also "patted some clay and set down the words as on a tablet . . . it verily was so." And so the first written words came into being, although the Sumerians sidestepped the awkward matter of who was able to read the lord of Kulab's words.

Less uncertain is the place of the Sumerian writing system as a milestone in the evolution of writing. It was a true system, with all that this implies for emerging cognitive skills in writer, reader, and teacher. Although far more comprehensive than tokens, the earliest signs in Sumerian cuneiform demanded only slightly more abstraction because they were generally pictographic (images that visually resembled the object represented). The pictographic characters were easily recognized by the visual system, which would require only a further match with an object name in the spoken language. Stanislas Dehaene observes that many of the symbols and letters used in writing and numerical systems around the world incorporate highly common visual shapes and features that correspond to objects in nature and our world. The French novelist Victor Hugo, quoted above, observed as much at the turn of the twentieth century. Hugo proposed that all letters originated from Egyptian hieroglyphs, which, in turn, stemmed from images in our world such as rivers, snakes, and lily stalks. These similar ideas between novelist and neuroscientists remain conjectural but they highlight the question of how the brain learned to recognize letters and words in the first place with such alacrity. In Dehaene's evolutionary terms, early pictographic symbols, which utilized known shapes in the external world, "recycled" the circuits used for object recognition and naming.

This simple state of affairs didn't last long, however. Soon after it originated, Sumerian cuneiform, mysteriously and rather astonishingly, became sophisticated. Symbols rapidly became less pictographic and more logographic and abstract. A logographic writing system directly conveys the concepts in the oral language, rather than the sounds in the words. Over time many of the Sumerian characters also began to represent some of the syllables in oral Sumerian. This double function in a writing system is classified by linguists as a *logosyllabary*, and it makes a great many more demands on the brain.

In fact, to fulfill these double functions, the circuits of the Sumerian reading brain must have crisscrossed it. First, considerably more pathways in the visual and visual association regions would be necessary in order to decode what would eventually become hundreds of cuneiform characters. Making such accomodations in the visual areas is basically the equivalent of adding memory to our hard drive. Second, the conceptual demands of a logosyllabary would inevitably involve more cognitive systems, which, in turn, would require more connections to visual areas in the occipital lobes, to language areas in the temporal lobes, and to frontal lobes. The frontal lobes become involved because of their role in "executive skills" such as analysis, planning, and focused attention, all of which are necessary to process the tiny syllables and sounds within words and the many semantic categories like human, plant, and temple.

Attending to individual sound patterns inside words must have been very new for our ancestors, and it came about because of something extremely clever. As they began to add new words, the Sumerians incorporated what is called a *rebus principle* in their writing. This occurs when a symbol (for example, "bird") represents not its meaning but rather its sound, which in Sumerian was a word's first syllable. In this way, the symbol for "bird" could do double duty—as its meaning or its speech sound. Disambiguating the two, of course, required still more new functions, including specific markers both for sounds and for common categories of meaning. These phonetic and semantic markers, in turn, required more elaborate cerebral circuitry.

To imagine what the Sumerian brain eventually looked like, we can do two tricky things. First, we can return to the findings of

Raichle's group, who looked at what happens when meaning is added to words. For example, they studied how the brain reads pseudo words like "mbli" and real words like "limb," in which the letters were the same but only one combination of them was meaningful. In each case, the same visual areas initially activated, but the pseudo words stimulated little activity beyond their identification in the visual association regions. For real words, however, the brain became a beehive of activity. A network of processes went to work: the visual and visual association areas responded to visual patterns (or representations); frontal, temporal, and parietal areas provided information about the smallest sounds in words, called phonemes; and finally areas in the temporal and parietal lobes processed meanings, functions, and connections to other real words. The difference between the two arrangements of the same letters—only one of which was a word—was almost half a cortex. When encountering words written in cuneiform and hieroglyphs, the first readers—both the Sumerians and the Egyptians—undoubtedly used parts of these same regions, as they set about creating the first two writing systems.

As further evidence of this scenario, I have a second trick up my sleeve. To get another glimpse of the ancient Sumerian reading brain, we can extrapolate from a living, flourishing, similarly constructed writing system (i.e., logosyllabary). One language today has a similar history of shifts from pictographic symbols to logographic symbols, uses phonetic and semantic markers to help disambiguate its symbols, and has ample brain images: Chinese. John DeFrancis, a scholar of ancient languages and Chinese, classifies both Chinese and Sumerian as logosyllabic writing systems, with many similar elements, though of course also some dissimilar ones.

Thus a Chinese reading brain (Figure 2-5) offers a contemporary, fairly reasonable approximation of the brains of the first Sumerian readers. A vastly expanded circuit replaces the little circuit system of the token reader. This new adaptation by the brain requires far more surface area in visual and visual association regions, and in both hemispheres. Unlike other writing systems (such as alphabets), Sumerian and Chinese show considerable in-

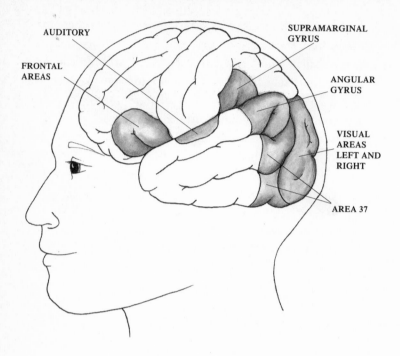

AUDITORY

FRONTAL
AREAS

SUPRAMARGINAL
GYRUS

ANGULAR
GYRUS

VISUAL
AREAS
LEFT AND
RIGHT

AREA 37

Figure 2-5: Logosyllabary—Reading Brain

volvement of the right hemisphere areas, known to contribute to
the many spatial analysis requirements in logographic symbols
and also to more global types of processing. The numerous, visually demanding logographic characters require much of both visual areas, as well as an important occipital-temporal region
called area 37, which is involved in object recognition and which
Deheane hypothesizes is the major seat of "neuronal recycling" in
literacy.

Although all reading makes use of some portions of the frontal and temporal lobes for planning and for analyzing sounds and
meanings in words, logographic systems appear to activate very
distinctive parts of the frontal and temporal areas, particularly
regions involved in motoric memory skills. The cognitive neuroscientists Li-Hai Tan and Charles Perfetti and their research group
at the University of Pittsburgh make the important point that
these motoric memory areas are far more activated in reading

Chinese than in reading other languages, because that is how Chinese symbols are learned by young readers—by writing, over and over. This is also how Sumerian characters were learned—on little clay practice tablets, over and over again: "it verily was so."

The Story within a Story: How Sumerians Taught their Children to Read

The Sumerians taught all new pupils to read with lists of words on little clay tablets. This small fact does not sound like a momentous event in the intellectual history of *Homo sapiens*, but it was. The act of teaching not only requires a firm knowledge of the subject, but also forces the teacher to analyze what goes into the learning of a particular content. Moreover, good teaching renders the multiple dimensions of the subject to be taught more visible—in this case, the complex nature of language in its written form. The gradual process of learning how to teach the earliest writing systems forced our world's first reading instructors to also become the world's earliest linguists.

Ancient records recently analyzed by the Assyriologist Yori Cohen of Tel Aviv University indicate how long it took Sumerian pupils before they could read and write—virtually years of study in their *e-dubba* or "tablet house" schools. This name refers to an essential part of the Sumerian instructional methods: teachers would write the cuneiform symbols on one side of a clay tablet, and students would copy the symbols on the reverse side. New readers learned to read text that included both logographic and phonetic information—sometimes in the same word. To do so, young readers had to have rich contextual background knowledge, well-honed automatic skills, and no small amount of cognitive flexibility to decide what value to give a particular written sign—logographic, phonetic-syllabic, or semantic—if they were to understand the texts at all fluently. This took years of practice. It is little wonder that newly discovered practice tablets depict miserable students in each year with their teacher, followed by the oft-repeated line "And then he caned me."

But frequent canings are not the real surprise. These first teachers of reading utilized highly analytical, linguistic principles for teaching that would be useful in any era. From early on, Cohen observed that novice readers learned lists of words based on one of several particular linguistic principles. Some lists taught *semantic*, or meaning-based, categories, with each category identified by specific markers. As the Sumerian writing system began to incorporate symbols for syllables, a second set of word lists was grouped on the basis of shared pronunciations. This meant that Sumerians were analyzing the sound-based or *phonological* system—the emphasis of most phonics-based reading programs today. In other words, long before twentieth-century educators would debate whether reading is best taught by phonics or by meaning-based methods, the Sumerians were incorporating elements of both in their early instruction.

A major contribution of early Sumerian writing is the way that teaching methods promoted conceptual development. Requiring Sumerian pupils or any children to learn semantically and phonetically related words helped them recall the words more efficiently, increase their vocabulary, and increase their conceptual knowledge. In current terms, the Sumerians used the first known metacognitive strategy to teach reading. That is, Sumerian teachers gave their pupils tools that made *explicit* how to learn something, and how to remember it.

Over time, the novice Sumerian readers also learned words that illustrated the common *morphological* properties of language (e.g., how two symbolic units can come together to make a new related word). Morphology is a system of rules for forming words from the smallest meaningful parts of a language, called morphemes. For example, in English the word "bears" is composed of two morphemes: the root word, "bear"; and the "s," which indicates either a plural noun or the present tense of the verb "to bear." Without this profoundly important capacity for combination in language, our vocabulary and conceptual possibilities would be severely narrowed, with dramatic implications for our intellectual evolution and for the cognitive differences between our primate cousins and ourselves.

The call system of one of our primate relatives, Nigerian putty-nosed monkeys, illustrates the importance of this type of combinatorial capacity in language. The putty-nosed monkey, like the vervet monkey, has two separate warning calls for its major predators. "Pyow" means that a leopard is nearby, and a "hacking sound" indicates the approach of the eagle. Recently, two Scottish zoologists observed that the monkeys have combined the two calls to make a new call, indicating to the young monkeys that it's time "to leave a site." Such an innovation among the putty-nosed monkeys is analogous to our use of morphemes to create new words, as the Sumerians frequently did in their writing system.

What is historically humbling about Sumerian writing and pedagogy is not their understanding of morphological principles, but their realization that the teaching of reading must begin with explicit attention to the principal characteristics of oral language. This is exactly what takes place today in the supposedly "cutting-edge" curricula in our own lab where we incorporate all major aspects of language in our reading instruction. It makes perfect sense. If you believe you are the first reading people on earth, and have no prior methods to influence how you teach, you try to figure out all the characteristics of your oral language in order to create a written version. For the first Sumerian teachers this resulted in a long-lasting set of linguistic principles that facilitated teaching and learning and also accelerated the development of cognitive and linguistic skills in literate Sumerians. Thus, with the Sumerians' contributions to teaching our species to read and write, the story began of how the reading brain changed the way we all think.

All of us. One of the less known but felicitous aspects of the Sumerian legacy has to do with the discovery that the women of the royal houses learned to read. Women possessed their own dialect, called Emesal, or the "fine tongue," as distinguished from the standard dialect Emegir, the "princely tongue." The feminine dialect differed in the pronunciation of many of its own separate words. We can only imagine the cognitive complexity required by pupils who had to switch dialects between passages where the goddesses spoke the "fine tongue" and the gods used their "princely" one. It is beautiful testimony to this ancient culture that some of the world's first recorded love songs and lullabies were composed by their women:

> *Come sleep, come sleep*
> *Come to my son*
> *Hurry sleep to my son*
> *Put to sleep his restless eyes,*
> *Put your hand on his sparkling eyes,*
> *And as for his babbling tongue*
> *Let not the babbling hold back his sleep.*

From Sumerian to Akkadian

It is also a testimony to the Sumerian writing system that at least fifteen peoples, including the early Persians and Hittites, adopted the Sumerian cuneiform script and the related teaching methods long after Sumerian ceased to be used. Just as cultures die out, so, too, do languages. By the beginning of the second millennium BCE, Sumerian was dying as a spoken language, and new readers began to learn "bilingual lists" for words in the increasingly dominant Akkadian language. By 1600 BCE, no speakers of Sumerian remained. All the more impressive, therefore, is the fact that the Akkadian writing system and its teaching methods continued to preserve many Sumerian written symbols and methods. Sumerian learning methods contributed to the educational process throughout Mesopotamian history. Indeed, there are memorable scenes as late as 700 BCE showing two scribes working intently side by side, one on clay tablets and one on papyrus—one with the ancients' writing, and one with the new.

Only around 600 BCE did Sumerian writing disappear. And even then, its impact remained, inside some of the characters and within the methods of Akkadian, the lingua franca from the third to the first millennium BCE. Akkadian became the language used and adapted by most of the peoples of Mesopotamia for some of the most important ancient documents in recorded history, beginning with the timeless descriptions of the human condition in the Akkadian *Epic of Gilgamesh*:

> For *whom have I labored? For whom have I journeyed?*
> For *whom have I suffered? I have gained absolutely Nothing for myself.*

Discovered in twelve stone tablets in the Nineveh library of Ashurbanipal, king of Assyria between 668 and 627 BCE, the *Epic of Gilgamesh* bears the name of Shin-eq-unninni, one of the first known authors in history. In this epic, which undoubtedly has motifs from far earlier oral legends, the hero Gilgamesh battles terrible foes, overcomes horrific obstacles, loses his beloved friend, and learns that no one, including himself, can escape the ultimate enemy of all humans—mortality.

Gilgamesh and the flurry of Akkadian writing that followed exemplify several important changes in the history of writing. The sheer volume of writing and the flowering of literary genres contributed hugely to the knowledge base of the second millennium BCE. The titles of these works tell their own story—from touching didactic texts like *Advice of a Father to His Son* and spiritual works like *Dialogue of a Man with His God* to mythic legends like *Enlil and Nilil*. The impulse to codify led to what is probably the first encyclopedia, modestly titled *All Things Known about the Universe*. Similarly, the *Code of Hammurabi* in 1800 BCE gave the world a brilliant codification of the laws of society under this great ruler, and the *Treatise of Medical Diagnostics and Prognostics* classified all known medical writings. The level of conceptual development, organization, abstraction, and creativity in Akkadian writing inevitably shifts any previous focus on what is cognitively required by an individual writing system to what aspects of cognitive development are being advanced.

Some features of Akkadian made it somewhat easier to use, with a caveat. Ancient languages like Akkadian, and other languages like Japanese and Cherokee, have a rather simple, tidy syllabic structure. Such oral languages lend themselves well to the type of writing system called a syllabary, in which each syllable, rather than each sound, is denoted by a symbol. (For example, when the Native American leader Sequoya decided to invent a writing system, he used a syllabary, a system well suited for the eighty-six syllables of Cherokee.) A perfectly rendered, "pure" syllabary for Akkadian, however, would have meant giving up the old Sumerian logograms and their ties with the past—something unacceptable to the Akkadians. Over time a linguistic compromise emerged, often used in other languages. The Akkadian writing system retained some of the old Sumerian logograms for common, important words like "king," but rendered its other words in the syllabary. In this way the ancient Sumerian language and culture survived—a matter of great pride to the Akkadian culture—even though the resulting writing system became more complex. Underlying most convoluted writing systems around the world can be found the wish of one culture to preserve a previous culture or language that shaped it.

The English language is a similar historical mishmash of homage and pragmatism. We include Greek, Latin, French, Old English, and many other roots, at a cost known to every first- and second-grader. Linguists classify English as a *morphophonemic* writing system because it represents both morphemes (units of meaning) and phonemes (units of sound) in its spelling, a major source of bewilderment to many new readers if they don't understand the historical reasons. To illustrate the morphophonemic principle in English, the linguists Noam Chomsky and Carol Chomsky use words like "muscle" to teach the way our words carry an entire history within them—not unlike the Sumerian roots inside Akkadian words. For example, the silent "c" in "muscle" may seem unnecessary, but in fact it visibly connects the word to its origin, the Latin root *musculus*, from which we have such kindred words as "muscular" and "musculature." In the latter words the "c" is pronounced and represents the *phonemic* aspect of our alphabet. The silent "c" of "muscle," therefore, visually conveys the *morphemic* aspect of English. In essence, English represents a "trade-off" between depicting the individual sounds of the oral language and showing the roots of its words.

An intellectually and physiologically demanding writing system confronted young readers of ancient Akkadian, because of similar trade-offs. It is hardly a surprise that the Akkadian writing system, like the earlier Sumerian system, took between six and seven years to master. This length of time and powerful political factors restricted literacy to a small, exclusive group of people in the temple and court—those who could afford the luxury of learning something for several years. Nowhere are such political forces lived out more vividly or more disastrously than in the parallel story of the other "first" writing system—the Egyptian hieroglyphs, which some recent scholarship suggests predated the Sumerian system by perhaps a century or more.

Another "First"? The Invention of Hieroglyphic Writing

For many years most scholars have assumed that the Sumerians invented the first system for recording language, and that the Egyptian script was partially derived from the Sumerian writing system. New linguistic evidence, however, suggests that an entirely independent invention of writing in Egypt took place either around 3100 BCE; or, on the basis of still controversial evidence from German Egyptologists in Abydos, as early as 3400 BCE— earlier than the Sumerian script. If this finding proves correct, hieroglyphs would be the first major adaptation in the evolution of the reading brain.

Because the evidence is not yet certain, I want to present the Egyptian hieroglyphic system (Figure 2-6) as a separate adaptation. Largely logographic and aesthetically beautiful, the earliest hieroglyphs were visually very unlike the Sumerian bird-feet style. Anyone who has tried to decipher some of this early writing soon becomes enamored of its sheer artistry. Both scripts employed the unusual rebus principle, and both were considered gifts from the gods.

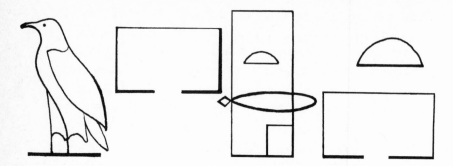

Figure 2-6: Egyptian Hieroglyphs of Bird, House, and Temple

Over time the hieroglyphic script evolved into a mixed system with both logographic signs for a core of word meanings and also special signs for consonantal sounds (called phonograms). For example, the hieroglyphic sign for "house" looks a lot like a house seen from above—as the gods were thought to see it. This sign can be used as a simple, imagistic logogram meaning "house," or it can be read as the consonantal blend "pr." Or it can be placed after other logograms to ensure that those signs are pronounced with "pr." This is the *phonetic marker* or *complement*, seen also in Sumerian. Or the sign can be placed with semantically similar, related words like "temple" and "palace" to ensure that the reader knows the classification of the word (see above).

With regard to cognitive requirements, the Egyptian system, like Sumerian, must have presented a formidable challenge to the novice reader. The early readers had to figure out exactly how a given sign was being used. Once again, the variety of strategies required by these different uses, combined with the cognitive judgment and flexibility involved in deciding when to use what, makes for a very active brain. To recognize a logogram required visual-conceptual connections; to recognize consonantal signs required connections between the visual, auditory, and phonological systems; and to recognize phonetic and semantic markers required additional abstraction and classification capacities, along with phonological and semantic analysis.

Furthermore, early Egyptian writing appeared neither punctuated nor consistently arranged from left to right or right to left. Egyptian and some other early systems were written in the *boustrophedon* style (Greek for "the turning around of the ox") in which one line moves from left to right and then right to left, the way oxen plow a field. Instead of scanning in one linear direction as we do today, the eye just moves down a notch at the end of a line and continues to read in the other direction. Egyptians also wrote from top to bottom or vice versa, depending on the architecture of the structure they were inscribing. The upshot is that the reader of hieroglyphs had to possess a spectrum of skills, including a highly developed visual memory, auditory and phoneme analysis, and considerable cognitive and spatial flexibility.

Over the centuries, like the Sumerian system and most other major early orthographies, the Egyptian system added many new signs and some new features. Unlike other systems, however, the Egyptian underwent two major transformations. First, for those charged with writing and copying, the hieroglyphic system evolved to include two cursive forms of writing. This first transformation added efficiency to the act of writing and copying texts, something which must have delighted all scribes. These ancient scribes, however, must have been even more pleased with the second transformation.

Basically, the Egyptians discovered the equivalent of the phoneme. There may have been no dancing in the streets, but for scribes this invention was very important indeed, for it helped them more easily denote new names of cities and members of the royal family, and also to spell foreign words and names. The clever rebus principle could take this task only so far. A similar phenomenon can be seen much later in Japanese's two writing systems—its older Chinese-based, logographic system, kanji; and the later syllabary, kana. (Like the Egyptian partial alphabet, the kana syllabary was designed as a supplement to kanji, to enable the written language to record new words, foreign words, and names.)

We know that this linguistic discovery was made in early Egyptian writing because it began to incorporate a tiny subset of characters that could depict the consonants of the Egyptians' oral language. As the linguist Peter Daniels described it, this was a marvel in the history of writing—the birth of a "partial *alphabet for consonants.*" This new group of characters by the Egyptians marked the earliest glimmerings of what would later become the third cognitive breakthrough in writing history: a system of writing based on the internal, sound-based structure of words. But just as Moses would be unable to live in the promised land, the Egyptians themselves would never fully exploit this alphabetic precursor. For reasons cultural, political, and religious, the hieroglyphic system never evolved in more efficient directions, despite the possibilities given to it by the partial alphabet. From some 700 standard signs in the Middle Egyptian period, the number of Egyptian hieroglyphs grew over the next millennium to several thousand, some of which became weighed down with layer after layer of religiously encrypted meanings, learnable by fewer and fewer people. These changes meant that hieroglyphic reading grew more, rather than less, conceptually demanding, and became restricted to fewer and fewer people.

We know from millions of Chinese readers, who daily acquire fluent reading also using thousands of characters, that the decline and fall of the Egyptian writing system cannot be explained simply by the quantitative demands on visual memory. By the first millennium BCE, the brain of an Egyptian scribe may well have needed far more cortical activation and cognitive resources to handle the encrypted meanings than was required for most other writing systems in all of history. Paradoxically, the Egyptian partial consonantal system—which may have first come into being because of the complexities of hieroglyphs—might prove to be the single most important contribution to the evolution of the alphabet in the early history of writing systems.

Dragon Bones, Turtle Shells, and Knots: The Curious Signs of Other Early Writing Systems

The very different histories of the Egyptian and the Sumerian systems do not resolve whether writing was invented separately by each culture, or whether one system traveled to the other. Cumulative evidence around the world suggests that writing was invented at least three times in the last part of the fourth millennium BCE, and at least three more times in different parts of the world in later periods. In addition to the Egyptian and Sumerian systems, the Indus people's system of writing evolved from potters' marks around 3300 BCE to a full script around 2500 BCE. This script remains undeciphered, and continues to defy valiant efforts to crack its code.

The first of the later writing systems appeared in Crete, in the second millennium BCE. Presumably influenced by the Egyptians, it included a pictographic, Cretan hieroglyphic script called Linear A, and the famous script Linear B. (See Chapter 3 on the Greek system.) A very different, rich logosyllabary system created originally by the Zapotecs was used by them, the Mayans, and the Olmecs throughout Mesoamerica. For decades, the stunning Mayan writing system, like the Greek Linear B, defied every attempt to decipher it. Then, quite remarkably, a relatively isolated scholar in Stalinist Russia with little access to most of the relevant materials broke the seemingly uncrackable code. Told in superb detail by Michael Coe in his book *Breaking the Maya Code*, the little-known story of Yuri Valentinovich Knorosov's breakthrough is one of the riveting intellectual whodunits of the twentieth century. Knorosov figured out that the brilliant ancient Mayans applied linguistic principles such as phonetic and semantic markers that were similar to those of the Sumerians and Egyptians, but that were even more similar to the way Japanese combined its two types of logographic and syllabary systems.

Another great Mesoamerican mystery, however, is still on the horizon. Recently, Gary Urton, an anthropologist at Harvard University, and his colleague Jeffrey Quilter suggested a new way of understanding the beautiful, mysterious quipus (or khipus), the ancient dyed fibers and twine shaped into patterns with extremely intricate systems of knots and attachments (refer back to Figure 2-1). Urton surprised linguists and Inca scholars with his hypothesis that the 600 or so quipus that still exist represent an undeciphered Incan written language system. Each type of knot, each direction of the knot, and each color may denote linguistic information, just as each knot in the Jewish tallith, or shawl, does. Until now, quipus were thought to have functioned like an abacus, although some records from Spanish historians in the sixteenth century described how the Incas told missionaries that entire cultures were recorded on them. (The missionaries promptly burned all the quipus they could find, to rid the Incas of their ties to past gods!) Today, Urton and his colleagues are trying to use the remaining quipus to decipher what may well be the equivalent of another complex ancient written language.

Yet another mystery can be found in the ancient Chinese writing system. Although its beginnings are usually dated from the Shang period (1500–1200 BCE), some scholars believe that a Chinese writing system existed much earlier. Another example of serendipity is the discovery of early Chinese writing in, of all places, nineteenth-century pharmacies. At the time people clamored to buy "dragon bones," which were believed to have magical healing properties, and someone noticed a system of marks on the old bones and the turtle shells. It is now thought that questions for the deities were written in an early Chinese script on turtle shells and the shoulder bones of cows; then the shells were split with a hot poker to reveal the gods' answers, given through the patterns of cracks that appeared inside the shells. A complete oracle bone inscription would ask the question, give the date, describe the gods' answers, and then tell what happened. For example, a 3,000-year-old inscription from the Shang dynasty recounts that King Wu Ding wanted to know whether his wife's pregnancy would be a "happy event." The gods answered that it would be happy only if the wife, Hao, gave birth on certain dates. She did not. The last inscription confirmed the prognostication of the deities: "The birthing was not a happy event. It was a girl."

Exquisitely formed characters hidden for centuries in turtle shells are a fitting metaphor for many Chinese characters, which contain whole histories within themselves. As we saw, like Sumerian, the Chinese system is a mixed logosyllabary that incorporates much of its past in its characters. As a result, it requires new readers to develop a prodigious amount of visual-spatial memory, which is enhanced by the act of writing these characters over and over. Just as for Sumerian and Egyptian phonetic complements, a small marker accompanies many of the most common characters to give information about a syllable's pronunciation. This sound-based feature helps distinguish some of the characters, whose visual features are otherwise difficult to learn and differentiate.

However, there are several ways in which Chinese differs from other ancient writing systems. First, it still exists. The Chinese writing system is a gift from the past to the present and is clearly hallowed by its readers. When Gish Jen, the celebrated Chinese-American novelist, traveled to China for a long stay, she noticed a very old man who came to a park every day with a long stick. Slowly over the course of an afternoon he would draw huge Chinese characters in the dry soil, each character perfectly rendered. The characters would be erased by the wind, but not before being admired by the people in the park. This scene captures the powerful ways that Chinese orthography incorporates not only a system for communication but also an artistic medium and, perhaps for this old Chinese man, an expression of spirituality as well.

In my own graduate seminar, I discovered another area of difference between the other ancient orthographies and Chinese. When I asked my Chinese students at Tufts University how they had learned so many characters at such a young age, they laughed and said they had a "secret system"—pinyin. Beginning readers learn pinyin to help them grasp the concept of reading and writing, to prepare them conceptually for having to learn 2,000 characters by the fifth grade. What is the secret of pinyin? It is a little alphabet. By giving young readers a sense of mastery over a small subset of characters, this Chinese alphabet prepares them to understand what reading is about and to tackle what lies ahead.

That's not the only surprise in Chinese. One of the loveliest ironies in the world's great mix of writing systems involves a very old Chinese writing system used only by women. Unlike the rest of Chinese writing, which is logographic, this system was completely based on phonetic translations of the sounds of Chinese words. The strange and wonderful story of *nu shu* writing— "female writing"—is poignantly depicted in Lisa See's novel *Snow*

Flower and the Secret Fan. Nu shu was drawn on delicately painted fans or sewn into beautiful fabrics in ritual letters. For centuries this remarkable writing system helped a small group of women endure and possibly transcend the constraints of lives symbolized by their tiny, bound feet. The last speaker and reader of *nu shu*, Yang Ituanyi, died recently at age ninety-six. *Nu shu* is a poignant reminder of the powerful role of writing in lives that might otherwise be isolated.

Nu shu also provides both an example of the wondrous diversity of the world's writing systems and a segue into the more phonetic-based systems, syllabaries, and alphabets. Like Chinese, alphabetic writing systems hide many mysteries, questions, and surprises. It is as if in trying to discover how many of us became readers of alphabets, we seek to learn something missing to us, something that we have always half known but that remains just out of reach. For Socrates it would have been just as well had none of us ever learned it, for reasons that should give us pause two and half millennia later.

Chapter 3

THE BIRTH OF AN ALPHABET
AND SOCRATES' PROTESTS

*There is a land called Crete . . . ringed by the wine-dark sea with
rolling whitecaps—handsome country, fertile, thronged with
people well past counting—boasting ninety cities,
language mixing with language side by side.*
—HOMER, *Odyssey*

Those who can read see twice as well.
—MENANDER (FOURTH CENTURY BCE)

ONE OF THE MOST INTRIGUING RECENT FINDINGS
in the history of written language took place in Egypt,
in Wadi el-Hol—which translates ominously as "the gulch of terror." In this desolate place where soaring, merciless heat scorches
the earth, the Egyptologists John Darnell and Deborah Darnell
found strange written inscriptions that predate by several centuries the earliest known alphabet. The script bore all the marks of
a "missing link" between the tiny Egyptian precursor system and
the later, beautiful Ugaritic script, which many scholars classify
as alphabetic. The Darnells believe that the Semitic scribes and
workers living in Egypt in the Hyksos period, around 1900–1800
BCE, invented this script, which appears to have exploited the

capacities of the small Egyptian consonantal system (to be expected) and to have within it some elements of the later Semitic
Ugarit script (not at all to be expected).

On examining the script from Wadi el-Hol, the renowned
scholar Frank Moore-Cross of Harvard University concluded that
this system is "clearly the oldest of alphabetic writing." He found
many symbols similar or identical to later known letters, and he
suggested that it "belongs to a single evolution of the alphabet."
The mysterious script of Wadi el-Hol is important because it con-
centrates our attention on the first of two multidimensional questions about a new adaptation of the brain for reading. First, what
makes up an alphabet and what separates it from the vestiges of a
previous syllabary, or logosyllabary? Responses to that question
prepare us to explore the second, larger question: are there significant intellectual resources unique to the alphabet-reading brain?

Wadi el-Hol's ancient script may well be a linguistic missing
link, connecting two types of writing systems—the syllabary and
the alphabet—but the dearth of available writings in this script
makes a thorough analysis difficult. The slightly later Ugaritic
system is a better-known candidate for the first alphabet and has
also been classified as both syllabary and an alphabet. The Ugaritic system originated in the rich and diverse coastal kingdom of
Ugarit (now the northern coast of Syria), an area bustling with
trade from both ships and overland caravans, all of which promoted the keeping of records. In Ugarit, different peoples spoke
at least ten languages, and five scripts could be found in addition
to its own. More important, the people of Ugarit left behind a
significant corpus of writings remarkable because it exhibits several key contributions of an alphabetic system. One such contribution is the economy that came with a reduced number of
symbols in its script.

Although Akkadian cuneiform was the original impetus for
the Ugaritic script, no Akkadian scribe could have deciphered the
new Ugaritic system of thirty signs, twenty-seven of which were
used in religious texts. In this unusual cuneiform-like system, independent consonant signs were combined with consonant signs
that distinguished adjacent vowels. Under the linguist William

Daniels's classification of writing systems, Ugaritic writing would be considered an *abjad*, or one particular type of alphabet, but this is a matter of debate.

However classified, the Ugaritic writing system represents a stunning accomplishment. It appeared in a range of genres from administrative documents to hymns, myths, and poems, and especially in religious texts. One of the most thought-provoking issues about it concerns the extent to which the oral and written Ugaritic language influenced the writing of the Hebrew Bible. A few scholars, including the biblical scholar James Kugel of Harvard, underscore the numerous similarities to the Old Testament in themes, in images, and in the often lyrical phrasing.

Another surprising discovery about the Ugaritic script involves its use of an "abecedary," as linguists call any system that lists the letters of a script in a fixed order. A curious item in the history of writing is that the same sequence in the Ugaritic abecedary characterized a second-millennium proto-Canaanite script, which went on to become the Phoenician consonantal system, which went on to become the Greek alphabet—or so the widely accepted account goes. Thus the abecedary is evidence of a link between these two candidates for early alphabets and also suggests some early schooling system that standardized the learning of letters in a fixed order. Like the Sumerians' use of lists, such an ordering gives novice readers a cognitive strategy for more easily remembering the characters of their script. But the use of this fascinating script ended when invaders destroyed Ugarit around 1200 BCE. Because of the disappearance of Ugarit, many questions remain unanswered about the ancient, beautiful written language that may well have helped fashion the evocative language of the Bible and that might have been one of our first functioning alphabetic systems.

There is a relevant biblically inspired short story by Thomas Mann about the creation of the alphabet. In this story, "The Law," God asks Moses to carve two tablets of stone, with five laws per tablet, that could be understood by all people. But how, Moses worries, is he supposed to write the words down? Moses knows the exotic letters of Egypt. He has seen the scripts of Mediterranean people, with signs resembling eyes, beetles, horns, and crosses. He also knows the syllable-writing of the desert tribes. But none of these signs for words and things can communicate the ten laws of God to everyone. In a burst of inspiration, Moses realizes that he must invent a universal system that people speaking any language could use to read their own words. And so he invents a form of writing in which every sound can have its own symbol and which all peoples can use to read their own languages: the alphabet. Using this new invention, Moses takes down God's dictated words and carves the whole in stone on Mount Sinai, not that far from Wadi el-Hol.

Although Mann was not a linguist or an archaeologist, he essentially described some of the revolutionary contributions of the alphabet and the core principles of the third cognitive breakthrough in the history of writing: the development of a writing system that requires only a limited number of signs to convey the entire repertoire of sounds in a language. Through a reduction of the signs to be learned in their writing systems, both the Wadi el-Hol and the Ugaritic scripts achieve the advantages that come from cognitive efficiency and a more economical use of memory and effort in reading and writing.

Cognitive efficiency depends on the third great feature of the brain: the ability of its specialized regions to reach a speed that is almost automatic. The implications of cognitive automaticity for human intellectual development are potentially staggering. If we can recognize symbols at almost automatic speeds, we can allocate more time to mental processes that are continuously expanding when we read and write. The efficient reading brain, which took Sumerian, Akkadian, and Egyptian pupils years to develop, quite literally has more time to think.

The questions suggested by these earliest alphabet-like systems are complex: Does the reduction of signs lead to a unique form of cortical efficiency? Are special cognitive capacities released in the alphabet-reading brain? What are the implications if such potential resources can occur early in the development of the novice reader? The trail to the answers begins with re-confronting a basic question.

What Makes an Alphabet?

Scholars in several disciplines continue to spar over the major conditions for a "true alphabet," based on the definitions in their own fields. Long before the discovery of the Wadi el-Hol script, the classicist Eric Havelock stipulated three criteria: a limited number of letters or characters (the optimal range was between twenty and thirty characters), a comprehensive set of characters capable of conveying the minimal sound units of the language, and a complete correspondence between each phoneme in the language and each visual sign or letter.

On this basis, classicists insist that all the alphabet-like systems before the Greek alphabet fail to meet these conditions. The Semitic scripts did not depict vowels; indeed, marks for vowels in Hebrew did not appear until millennia later, when the languages spoken in everyday life (such as Aramaic and Greek) made the explicit depiction of vowels more important. For classicists like Havelock, the alphabet represented the apex of all writing; and the Greek system (750 BCE) was the first to satisfy all conditions for a true alphabet, and the first that allowed huge leaps in humans' powers of thought.

Many linguists and scholars of ancient languages differ dramatically from this view. The Assyriologist Yori Cohen stresses something not discussed by Havelock. He views an alphabet as a system that uses the minimum of notations necessary to express a spoken language unambiguously to its native speakers. To Cohen, any system that can denote the smallest segments of sounds analyzable by the human ear in an oral language—as opposed to larger segments such as syllables or whole words—can be considered an alphabet. According to this view, the Ugaritic script and possibly the earlier Wadi el-Hol script would be classified as early types of alphabets.

I can pull no rabbit out of a hat to resolve this question. No universal agreement exists about this important "first" in human history. A recent burgeoning of new information about ancient writings, however, may give twenty-first-century readers a different, meta-view. By tracing the systematic changes of cognitive and linguistic skills over the early history of different systems leading up to the Greek alphabet, we can get a fresh insight into the slow transition from the oral world of Homer, Hesiod, and Odysseus on the shores of Cephalonia, Ithaca, and Crete to the changing Athenian worlds of Socrates, Plato, and Aristotle. The changes occurred not only in place and time but in memory and in the human brain itself. The next major adaptation of the reading brain was about to emerge.

THE MYSTERIOUS WRITINGS OF CRETE AND THE DARK AGES OF GREECE

On the island of Crete there is a myth to be found under every stone, but the simple truth is fascinating enough. For example, the stone itself may be part of an ancient Minoan civilization— perhaps a remnant of one of the exquisite frescoed royal palaces, where early forms of plumbing and air-conditioning were de rigueur. Four millennia ago Minoans built monuments and made art and jewelry of incomparable beauty, and they created systems of writing that continue to frustrate our best efforts at decipherment.

In 1900 Arthur Evans, a British archaeologist, dug up the ancient center of Minoan culture, Homer's great city of Knossos. This was the fabled site of King Minos's palace, with its intrepid bull-leapers and its fearful labyrinth inhabited by the Minotaur. During this dig Evans made an extraordinary discovery that would become a lifelong obsession: 7,000 clay tablets covered with an undecipherable script. Resembling neither Egyptian hieroglyphs nor Akkadian cuneiform, the script had features of an earlier Cretan script, called Linear A, but appeared unrelated to the later Greek alphabet. Evans called it Linear B, and began forty frustrating years of trying to decipher it.

In 1936, a studious teenager named Michael Ventris met Evans and became equally obsessed with the strange script. In 1952, Ventris finally deciphered the odd writing. Despite having stumped scholars for half a century, Linear B turned out to be anything but mysterious; it was, put simply, a crude rendition of spoken Greek at that time. To Ventris's classically trained mind, the anticlimactic discovery may have felt like cracking the code for an ancient version of Instant Messenger. Ventris never expected to decipher colloquial Greek, but in the words of the esteemed classical scholar Steve Hirsh of Tufts University, his decipherment of Linear B "revolutionized our knowledge of early Greece."

We still know very little about Linear B beyond its beginnings in the fifteenth century BCE in Crete, mainland Greece, and Cyprus, and its disappearance between the twelfth and eighth centuries BCE. During this time, called the dark ages of Greece, invasions destroyed most palaces—the repositories of literacy— and few records remain. But in this supposedly dark period a highly developed oral culture nevertheless thrived, captured for all time in the work of Homer in the eighth century BCE. Whether Homer was the blind bard of myth (there is new reason to believe so) or several poets, or even the cumulative memory of an oral culture remains unresolved. What is undisputed is that the encyclopedic knowledge and mythology in Homer's *Iliad* and *Odyssey* contributed strongly to the formative development of every Greek citizen. According to the Greek historian Thucydides, an educated Greek citizen committed to memory huge passages of that epic history, with its poignant stories of Greek gods, goddesses, heroes, and heroines.

Certainly, as the great scholar of this period, Walter Ong, argued, many aspects of epic poetry lent themselves to memorization: the driving meter and the rich melodic quality of the highly rhythmic Homeric lines; the often-repeated, vivid images (such as "rose-fingered dawn"); and the very subjects of the *Iliad* and *Odyssey*, with their timeless stories of love, war, virtue, and frailty. The scholar Millman Perry, for example, found that multiple, well-known formulas describing stock accounts of various deeds and events were pieced together by the bards, generation after generation. Combined with the Greek orators' other renowned "mnemotechniques," these formulas enabled ancient Greeks to memorize and recite amounts of material that would intimidate most of us today. One of these fabled memory techniques required the person to associate physical spaces—like the interiors of imagined libraries and temples—with the things to be recalled.

The poet Simonides provides an astonishing, concrete example of the Greeks' legendary memory. Once, when an earthquake destroyed a building where he and many other people had been celebrating, he recollected the names of all the people who had attended and figured out exactly where they lay buried in the rubble.

How could Simonides and other Greeks achieve such powers of recall? During the last 40,000 years or so, all human beings have shared the same basic brain structure, so there is little reason to think that there were structural differences between ourselves and our Greek ancestors in the hippocampus, amygdala, frontal lobes, or other regions serving memory. What distinguished our ancestors in ancient Greece from us was the great value the Greeks placed on an oral culture and memory. Just as Socrates probed his students' understanding in dialogue after dialogue, educated Greeks honed their rhetorical and elocutionary skills, and prized above almost everything else the ability to wield spoken words with knowledge and power. The astounding memory capacities of our Greek ancestors are one result. They remind us of the significant effects of culture on the development of presumably innate cognitive processes, such as memory.

Into this highly developed oral culture, the Greek written alphabet stumbled headfirst. Several scholars suggest that the Greek written alphabet came into existence largely because the Greeks wanted to preserve the oral traditions of Homer—that is, the alphabet had a role subservient to oral language. Whatever the case, ancient Greeks would be flabbergasted to know that scholars today, 2,700 years later, remain in awe of their achievement—an achievement which would quietly diminish the use of their prized memory and rhetorical capacities and unleash new and different forms of memory and cognitive resources that continue to shape us in the present day.

The "Invention" of the Greek Alphabet: Daughter of the Phoenicians or Sister?

If ancient Greeks had been asked where their alphabet came from, they would probably have replied that they just borrowed it. They called their alphabet "Phoenician letters," reinforcing the belief that its most direct ancestor was the Phoenician consonant-based script. The Phoenicians, in turn, based their letters on earlier Canaanite scripts. (And indeed the Phoenicians referred to themselves as Canaanites.) The Greek letters *alpha* and *beta* come from the Phoenician *aleph* and *bet*, other evidence of Phoenician roots. Recent scholarship, however, finds no such neat lineage. At least one quiet war is being waged over different constructions of how the Greek alphabet developed.

The first construction is what the German scholar Joseph Tropper calls the "standard theory" of the alphabet's origins: the Greek alphabet stems from Phoenician, which stems from an earlier Ugaritic or proto-Canaanite script, which possibly stems from the small set of Egyptian consonant-based characters. However, another German scholar, Karl-Thomas Zauzich, strongly asserts a different interpretation of the evidence: "Greek writing is not a daughter of Phoenician writing but the sister! These two writing systems must have had a shared Semitic mother, from whom no witnesses now can speak." Zauzich argues that Greek writing resembles the original Egyptian cursive script far more than Phoenician does. From this and other evidence, he concludes that the Greek alphabet was not a Phoenician offshoot at all, but rather a coequal descendant from a shared earlier system: a sister, as he puts it.

Mythology is tricky source material. That said, according to no small number of myths, the alphabet came to Greece from Cadmus (in Greek, Kadmos), the legendary founder of Thebes, whose name means "east" in Semitic. This may indicate that some Greeks were aware of the Semitic origins of their alphabet. Whatever the intention, Greek myths about how the gods gave letters to the mortal Cadmus rival the tales of the Grimm brothers for gore; at least one version ends with Cadmus strewing bloodied teeth (metaphorical letters) into the ground to grow and spread.

Like these allegorical teeth, the drama of the Greek alphabet lies beneath the surface. The textbook account, which is similar to the "standard theory," goes like this. Between 800 and 750 BCE the Greeks designed their alphabet and disseminated it to Greek trading colonies in Crete, Thíra, El Minya, and Rhodes. To do this, the Greeks systematically analyzed each of the phonemes of

the Phoenician and the Greek languages. Then, using the Phoenician consonant-based system as a base, they created their own
symbols for vowels, doggedly perfecting the correspondence between letters and all known sounds. On this basis, the Greek alphabet became the progenitor of most Indo-European alphabets
and systems, from Etruscan to Turkish. Below those details lies a
series of mysteries for cognitive scientists and linguists, beginning
with the second overarching question of this chapter.

Does an Alphabet Build a Different Brain?

Whenever people or humanlike creatures come together (see Dr.
Seuss's tale *The Sneetches*), some group at some time claims superiority. So too with writing. Various influential twentieth-century
scholars have argued that the alphabet represents the apex of all
writing and that, consequently, alphabet readers "think differently." In the context of our cognitive history, three claims about
supposedly unique contributions of the alphabet now lend themselves to our analysis: (1) the alphabet's increased efficiency over
other systems; (2) the alphabet's facilitation of novel thoughts,
never before articulated; and (3) the novice readers' ease in acquiring an alphabetic system through their increased awareness
of the sounds of speech. (This ability enables children to hear and
analyze phonemes; thus it facilitates learning to read and helps
spread literacy.)

CLAIM 1: THE ALPHABET IS MORE EFFICIENT
THAN ALL OTHER WRITING SYSTEMS

Efficiency is the capacity of a writing system to be read rapidly
with fluent comprehension. The alphabet achieves its high level
of efficiency through its economy of characters (a mere twenty-
six letters in many alphabets, compared with 900 cuneiform
characters and thousands of hieroglyphs). This reduced number
of symbols reduces the time and attention needed for rapid recognition; and thus fewer perceptual and memory resources are
needed.

In the history of writing that leads up to the alphabet, however, examining the brain can help us examine this claim. The remarkable rapidity and efficiency achieved by the Chinese, who must read thousands of characters, is on display in brain images of modern Chinese readers (see Figure 3-1). These images show the brain's vast capacity for visual specialization when both hemispheres are recruited in reading all of the many characters. The Chinese reader's fluency is one proof that efficiency is not reserved for alphabet readers alone. The syllabary reader's brain is another proof. Together they illustrate that more than one adaptation can lead to efficiency. They do not, however, address whether fluent reading in each type of system is equally achievable by most readers.

We can see the different types of efficiency among languages if we look at composite drawings of three reading brains in Figure 3-1. The alphabet-reading brain differs substantively from that of the earlier logosyllabary reader in the decreased amount of cortical space it needs in some areas. Specifically, the alphabet reader learns to rely more on the posterior of the left hemisphere, specialized regions with less bihemisphere activation in these visual regions. By contrast, the Chinese (and Sumerians) achieve efficiency by recruiting many areas for specialized, automatic processing across both hemispheres.

This differential use of hemispheres is clear in a fascinating early bilingual case study, written by three Chinese neurologists in the late 1930s. In their account of a bilingual person who suddenly developed alexia (lost the ability to read), they described how a businessman, proficient in Chinese and English, suffered a severe stroke in the posterior areas. What was amazing to all at the time was that this patient, who had lost his ability to read Chinese, could still read English.

Today, this example no longer seems bizarre, because current brain imaging shows us how the brain can be differentially organized for different writing systems. Japanese readers offer a particularly interesting example because each reader's brain must learn two very different writing systems: one of these is a very efficient syllabary (kana) used especially for foreign words, names of cities, names of persons, and newer words in Japanese; and the

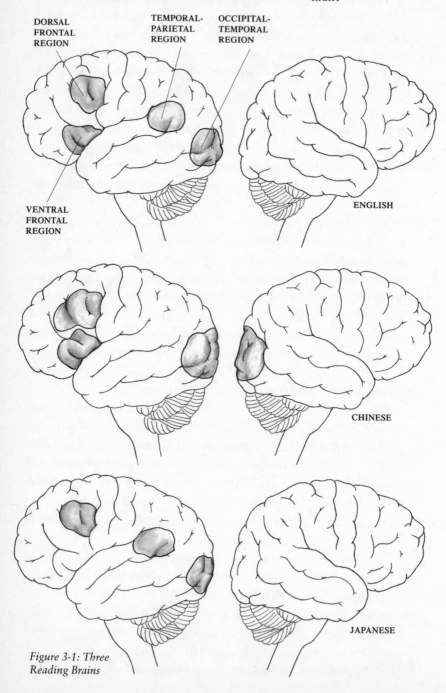

LEFT   RIGHT

DORSAL
FRONTAL
REGION

TEMPORAL-
PARIETAL
REGION

OCCIPITAL-
TEMPORAL
REGION

VENTRAL
FRONTAL
REGION

ENGLISH

CHINESE

JAPANESE

*Figure 3-1: Three
Reading Brains*

second is an older Chinese-influenced logographic script (kanji). When reading kanji, Japanese readers use pathways similar to those of the Chinese; when reading kana, they use pathways much more similar to alphabet readers. In other words, not only are different pathways utilized by readers of Chinese and English, but different routes can be used within the same brain for reading different types of scripts. And because of the brain's prodigious ability to adapt its design, the reader can become efficient in each language. Also, efficiency itself is not a binary, "either-or" operation. Japanese researchers find that the same words written in kana, its syllabary system, are read faster than kanji. Therefore, we can see that efficiency may be best conceptualized as a continuum, not the exclusive achievement of an alphabet.

That said, if we were able to look at all the ways the brain has learned to read over this early history, we would find some areas of great similarity and some features truly unique to each written language. In a pathbreaking meta-analysis of twenty-five imaging studies of different languages, cognitive scientists from the University of Pittsburgh found three great common regions used differentially across writing systems. In the first, the occipital-temporal area (which includes the hypothesized locus of "neuronal recycling" for literacy), we become proficient visual specialists in whatever script we read. In the second, the frontal region around Broca's area, we become specialists in two different ways—for phonemes in words and for their meanings. In the third, the multifunction region spanning the upper temporal lobes and the lower, adjacent parietal lobes, we recruit additional areas that help to process multiple elements of sounds and meanings, which are particularly important for alphabetic and syllabary systems.

Viewed alongside each other, these brain regions provide an emerging picture of what the cognitive scientist Charles Perfetti of the University of Pittsburgh and his colleagues call a "universal reading system." This system connects regions in the frontal, temporal-parietal, and occipital lobes—in other words, select areas from all four lobes of the brain.

A glance at these collective images helps us visualize two important conclusions about the evolution of writing: one, reading in any language rearranges the length and breadth of the brain; and two, there are multiple pathways to fluent comprehension, with a continuum of efficiency taking varied forms among the varied writing systems. Factors like the number of symbols in a writing system, the sound structure of an oral language, the degree of regularity in a written language, the degree of abstraction, and the extent of motoric involvement in learning a script will influence both the efficiency and the specific circuitry of a writing system. Together, they contribute to how easily the novice reader acquires reading. Indeed, not only are words in the kana syllabary read faster than the logographic kanji; children learning more regular alphabets, such as Greek and German, gain fluency and efficiency faster than children learning less regular alphabets, such as English.

Philosophers including Benjamin Whorf and Walter Benjamin have raised questions about whether different languages influence the minds of their individual readers in particular ways. The three claims I'm addressing about the alphabet are far more delimited, but there are differences to be found. As the neuroscientist Guinevere Eden at Georgetown observed, different writing systems set up their own distinctive brain networks in the development of reading. Within that narrowed context, the alphabet does not build a "better" brain, but it builds a brain different from the brain in other writing systems, in terms of its particular form of developmental efficiency.

More specifically, the "distinctive brain networks" developed earlier and more efficiently in the young Greek alphabet reader than in young readers of Sumerian or Egyptian. This is not to claim that developmental efficiency is unique to alphabetic systems. When the oral language is better represented by a syllabary—for example, in Japanese and Cherokee—a syllabary is equally efficient, in terms of both acquisition time and cortical space. The cortical efficiency gained from a smaller number of symbols—whether alphabet or syllabary—and the consequent developmental efficiency gained during their acquisition mark one of the great transitions in the history of writing. Whether cortical efficiency and developmental efficiency contribute far more than just speed moves us to the second major claim for the alphabet—novel thinking.

CLAIM 2: THE ALPHABET STIMULATES NOVEL THOUGHT BEST

The classicist Eric Havelock and the psychologist David Olson assert the thought-provoking hypothesis that the efficiency of the Greek alphabet led to an unparalleled transformation in the actual content of thought. By liberating people from the effort required by an oral tradition, the alphabet's efficiency "stimulated the thinking of *novel thought*."

Try to imagine a situation in which the educated members of an oral culture had to depend entirely on personal memorization and meta-cognitive strategies to preserve their collective knowledge. Such strategies, however impressive, came with a cost. Sometimes subtly, sometimes blatantly, dependence on rhythm, memory, formulas, and strategy constrained what could be said, remembered, and created.

The alphabet and other writing systems did away with most of those constraints, thereby enlarging the boundaries of what could be thought and written by more people. But is this a unique contribution of the Greek alphabet, or is it the very act of writing that promotes new levels of thought for more people? If we look back almost 1,000 years before the Greeks to the Ugaritic writing system, we can observe a good example of what any alphabet-like system can do within a culture. If we look back still earlier to Akkadian literature, which Havelock did not study, we see an outpouring of thought (some of it, to be sure, based on oral tradition), recorded by a nonalphabetic logosyllabary.

By taking a meta-view of this entire history, we can see that what promotes the development of intellectual thought in human history is not the first alphabet or even the best iteration of an alphabet but writing itself. As the twentieth-century Russian psychologist Lev Vygotsky said, the act of putting spoken words and unspoken thoughts into written words releases and, in the process, changes the thoughts themselves. As humans learned to use written language more and more precisely to convey their thoughts, their capacity for abstract thought and novel ideas accelerated.

Every child who learns to read someone else's thoughts and write his or her own repeats this cyclical, germinating relationship between written language and new thought, never before imagined. This generative relationship shines through the early history of writing from the Egyptian instructions on the afterlife through the Babylonian *Dialogues on Pessimism* to Plato's *Dialogues*. But within this history of writing, there is no debate that one of the best examples of the creative reciprocity between writing and thinking is the Greek alphabet.

From a cognitive perspective, therefore, it is again not that the alphabet uniquely contributed to the production of novel thought, but rather that the increased efficiency brought about by alphabetic and syllabary systems made novel thought more possible for more people, and at an earlier stage of the novice reader's development. This, then, marks the revolution in our intellectual history: the beginning democratization of the young reading brain. Within such a broadened context, there can be no surprise that one of the most profound and prolific periods of writing, art, philosophy, theater, and science in all of previously recorded history accompanied the spread of the Greek alphabet.

CLAIM 3: THE ALPHABET FACILITATES
READING ACQUISITION THROUGH ENHANCED
AWARENESS OF SPEECH

The Greek alphabet did differ dramatically from previous writing systems in its incorporation of highly sophisticated linguistic insights into human speech. The ancient Greeks discovered that the entire speech stream of oral language could be analyzed and systematically segmented into individual sounds. This is not an obvious perception for anyone, in any era. It is especially fitting that the Greeks, the most vocal proponents of oral culture, discovered for themselves the underlying structure and components of speech.

To understand the prodigious feat involved in the Greeks' analysis of speech, one has only to look to the Department of Defense! The modern history of speech perception begins with concentrated efforts to study the components of speech during World War II, when communication under the most challenging conditions was essential. This research started as a highly classified military secret, as scientists from Bell Laboratories tried to build machines that could analyze what is called the "speech signal" and ultimately synthesize human speech. When an entire battle could hinge on one officer's ability to hear a message in a bombarded trench in a war zone, such information was imperative for defense. Bell researchers used a new adaptation of an instrument called the spectrograph to observe a visual form of several critical components: the distribution of sound frequencies contained in an utterance; the duration or time required for each part of the signal; and the amplitude of a given signal. Each sound in every language has a signature made up of these three components.

As the acoustic properties of each aspect of human speech became more "visible" to modern researchers, the overwhelming complexity of speech became more visible. To take one small example, research by the speech scientist Grace Yeni-Komshian indicates that we speak at a rate of about 125 to 180 continuous words per minute, with no acoustic cues at the beginning or end of our words. (Think for a moment how an unknown foreign language sounds to you; it's a continuous, incomprehensible stream of sounds.) Within any language we speak, we know how to segment speech units by virtue of their meaning, grammatical roles, and morphological units, and by the cues provided by rhythm, stress, and intonation. Such information, however, helps very little in knowing where the first sound within a word (its onset) ends and the second sound begins. This is because all sounds are co-articulated, or "shingled together," with one phoneme overlapping with and dictating the pronunciation of the next. Yeni-Komshian wrote: "One of the greatest challenges for speech perception researchers is to determine how individual sounds are isolated (segmented) from the complex speech signal and how they are identified appropriately."

The Greek inventors of their alphabet did exactly that. First, as sketched in the textbook account, they systematically analyzed

each phoneme in Phoenician, along with the correspondence between these sounds and the Phoenician letters. Then, they did the
same analysis with Greek speech. Next, using the recycled Phoenician graphemes as a basis, they eventually matched almost every phoneme in the Greek language with a Greek letter; this
entailed making new letters for vowel sounds. For example, the
Greek alpha for the vowel "a" emerged from the Phoenician word
aleph, which means "ox." In one fascinating linguistic innovation, Greek writers changed some symbols to better match linguistic features of the dialect spoken in a given locale. This is
thought to be the reason why slightly differing scripts appear in
different Greek cities. Changing the letters of a writing system to
match a local dialect represents an ingenious stroke of linguistic
pragmatism and phonological expertise that would scarcely even
be considered today by members of the Académie Française. Only
when the staggering complexity of all speech is fully understood
can we appreciate what the Greeks did. If the Sumerians were the
first known general linguists, and the Sanskrit scholars were the
first grammarians, the Greeks were the first phoneticians.

The great breakthrough by the inventors of the Greek alphabet—the conscious, systematic analysis of speech—happens un-
consciously in the life of every child who learns to read. Young
Greek pupils were given an almost perfect alphabet with almost
perfect rules of grapheme-phoneme correspondence. As a result,
these pupils could gain fluency in literacy far sooner than their
Sumerian, Akkadian, or Egyptian counterparts. Beyond the scope
of this book, the question emerges whether this earlier development of fluency in the ancient Greek readers resulted in the expansion of thought that helped usher in the great classical Greek
period.

In the context of that unanswered question, a striking irony is
the centuries- long ambivalence of the Greeks toward teaching
the Greek alphabet. Shortly after the creation of their revolutionary script, the main reaction in Greece seems to have been a 400-
year thud. In stark contrast to the Egyptians and Akkadians,
educated Greeks considered their highly developed oral culture
superior to a written culture.

The historical figure of Socrates represents the most eloquent apologist for an oral culture and the most vigorous questioner of a written one. Before too quickly dismissing the ambivalence of the Greeks toward the invention of the Greek alphabet, we need to ask why one of the world's most accomplished thinkers and producers of novel thought decried its use. We turn now to an invisible war between an oral language culture and the use of written language in Greece. Plato's careful recording of Socrates' surprising arguments against literacy reveals tremendously important reasons why we would do well to heed them today.

Socrates' Protests, Plato's Quiet Rebellion, and Aristotle's Habit

Socrates himself wrote nothing at all. If we are to believe the account of his reasons given in Plato's Phaedrus, it was because he believed that books could short-circuit the work of active critical understanding, producing a pupil who has a "false conceit of wisdom."
—MARTHA NUSSBAUM

It is not too much to say that with Aristotle the Greek world passed through oral instruction to the habit of reading.
—SIR FREDERIC KENYON

He lived and dressed simply, and he described himself as the "stinging gadfly" on the back of the noble but sluggish horse that was Greece. With prominent eyes, bulging forehead, and little conventional physical beauty, he stood in a courtyard surrounded by students, involved in intense dialogue about abstract beauty, knowledge, and the profound importance of an "examined life." When he spoke, he possessed an extraordinary power as he exhorted the youth of Athens to devote themselves to a lifelong examination of "truth." This is the man we know as Socrates— philosopher, teacher, and citizen of Athens.

In writing the history of the early reading brain, I was surprised to realize that questions raised more than two millennia ago by Socrates about literacy address many concerns of the early twenty-first century. I came to see that Socrates' worries about the transition from an oral culture to a literate one and the risks it posed, especially for young people, mirrored my own concerns about the immersion of our children into a digital world. Like the ancient Greeks we are embarked on a powerfully important transition—in our case from a written culture to one that is more digital and visual.

I regard the fifth and fourth centuries BCE, when Socrates and Plato taught, as a window through which our culture can observe a different but no less remarkable culture making an uncertain transition from one dominant mode of communication to another. Few thinkers could be as capable of helping us examine the place of oral and written language in the twenty-first century as the "gadfly" and his pupils. Socrates passionately decried the uncontrolled spread of written language; Plato was ambivalent, but used it to record arguably the most important spoken dialogues in written history; and as a youth Aristotle was already immersed in "the habit of reading." These three figures are one of the world's most famous academic dynasties, for Socrates was mentor to Plato, who was mentor to Aristotle. Less known, if Plato's descriptions of Socrates' own history are factual, is that Socrates was the pupil of Diotima, a woman philosopher from Manitea, who used dialogues to teach her students.

Made immortal by Plato, the dialogues between Socrates and his students served as a model for what Socrates believed all Athenian citizens should do for their own growth as humans. Within these dialogues every pupil learned that only the examined word and the analyzed thought could lead to real virtue, and only true virtue could lead a society to justice and could lead individuals to their god. In other words, virtue, both in the individual and in society, depended on a profound examination of previous knowledge, and the internalization of its highest principles.

This intensive mode of learning differed radically from most previous Greek traditions, in which individuals received a collective wisdom handed down to them, exemplified by Homer's epics.

Socrates taught students to question the words and concepts conveyed through spoken language so they could see what beliefs and assumptions lay beneath them. Socrates demanded that everything be questioned—a passage from Homer, a political issue, a single word—until the essence of the originating thought became clear; understanding how it reflected—or failed to reflect—the deepest values of society was always the goal, and the questions and answers in dialogue were the vehicles of instruction.

Socrates was put on trial for corrupting youth with his teachings. Five hundred citizens of Athens declared his crime punishable by death. Some charged that he did not believe in the gods. To Socrates, such claims cloaked political efforts to punish him for maintaining friendships deemed dangerous to the state, and to curb his questioning of accepted wisdom. His death by poison is ultimately far less important than his lifelong example of examining "with all our intelligence" what we do, what we say, and what we think. His exhortations ring out across time, echoing in our ears many centuries later. Here is a passage from his famous speech at his trial:

> If I tell you that this is the greatest good for a human being, to engage every day in arguments about virtue and the other things you have heard me talk about, examining both myself and others, and if I tell you that the unexamined life is not worth living for a human being, you will be even less likely to believe what I am saying. But that's the way it is, gentlemen, as I claim, though it's not easy to convince you of it.

In examining written language, Socrates took a stand that usually comes as a surprise: he felt passionately that the written word posed serious risks to society. His three concerns appear disarmingly simple, but they are not. And as we examine our own intellectual transition to new modes of acquiring information, these objections deserve our every effort to get to their essence. First, Socrates posited that oral and written words play very different roles in an individual's intellectual life; second, he regarded the

new—and much less stringent—requirements that written language placed both on memory and on the internalization of
knowledge as catastrophic; and third, he passionately advocated
the unique role that oral language plays in the development of
morality and virtue in a society. In each instance Socrates judged
written words inferior to spoken words, for reasons that remain
powerfully cautionary to this day.

SOCRATES' FIRST OBJECTION: INFLEXIBILITY OF THE WRITTEN WORD

*The way of words, of knowing and loving words, is a way to
the essence of things, and to the essence of knowing.*
— JOHN DUNNE

In the film *The Paper Chase*, Charles Kingsfield, a professor of
law at Harvard, terrorizes his young students with his daily inter-
rogations. He demands that they justify whatever they say with
legal precedents. In the first classroom scene, Kingsfield declares,
"We use the Socratic method here . . . answering, questioning,
answering. Through my questions you learn to teach your-
selves. . . . At times you may think you have the answer. I assure
you this is a total delusion. In my classroom there is always another question. We do brain surgery here. My little questions are
probing your brain."

Kingsfield is a fictional embodiment both of the modern-
day Socratic method, and of a well-functioning reading brain.
Teachers and professors in many classrooms today continue this
probative function as they engage their students, analyzing the
assumptions and intellectual basis of every exchange. Such classroom scenes reenact variations of the critical inquiry once found
in Athenian courtyards. Professor Kingsfield demands that his
students know legal precedents, so that their understanding of
law can help preserve societal justice. Socrates wanted his students to know the essence of words, things, and thoughts, so that
they could acquire virtue, for it was virtue that led to "being called
the friend of God."

Underlying the Socratic method lies a particular view of words—as teeming, living things that can, with guidance, be linked to a search for truth, goodness, and virtue. Socrates believed that unlike the "dead discourse" of written speech, oral words, or "living speech," represented dynamic entities—full of meanings, sounds, melody, stress, intonation, and rhythms—ready to be uncovered layer by layer through examination and dialogue. By contrast, written words could not speak back. The inflexible muteness of written words doomed the dialogic process Socrates saw as the heart of education.

Few scholars would have been more comfortable with the importance Socrates gave "living speech" and the value of dialogue in the pursuit of development than Lev Vygotsky. In his classic work *Thought and Language*, Vygotsky described the intensely generative relationships between word and thought and between teacher and learner. Like Socrates, Vygotsky held that social interaction plays a pivotal role in developing a child's ever-deepening relationships between words and concepts.

But Vygotsky and contemporary scholars of language part ways with Socrates over his narrow vision of written language. In his brief life Vygotsky observed that the very process of writing one's thoughts leads individuals to refine those thoughts and to discover new ways of thinking. In this sense the process of writing can actually reenact within a single person the dialectic that Socrates described to Phaedrus. In other words, the writer's efforts to capture ideas with ever more precise written words contain within them an inner dialogue, which each of us who has struggled to articulate our thoughts knows from the experience of watching our ideas change shape through the sheer effort of writing. Socrates could never have experienced this dialogic capacity of written language, because writing was still too young. Had he lived only one generation later, he might have held a more generous view.

Hundreds of generations later, I wonder how Socrates might have responded to the capacity for dialogue in the interactive dimension of communication in the twenty-first century. The capacity of words to "speak back" is with us in many different ways, as children text-message each other, as we e-mail one another,

and as machines speak, read, and translate into different languages. Whether these capacities are being developed in ways that
sufficiently reflect the true, critical examination of thought would
be, for Socrates and for us, the essential question.

A more subtle concern for Socrates is that written words can
be mistaken for reality; their seeming impermeability masks their
essentially illusory nature. Because they "seem . . . as though they
were intelligent" and, therefore, closer to the reality of a thing,
words can delude people, Socrates feared, into a superficial, false
sense that they understand something when they have only just
begun to understand it. This would result in empty arrogance,
leading nowhere, contributing nothing. In this worry, Socrates
and Professor Kingsfield are bedfellows with thousands of teachers and parents today who watch their children spend endless
hours before computer screens, absorbing but not necessarily understanding all manner of information. Such partial learning
would be unthinkable to Socrates, for whom true knowledge, wisdom, and virtue were the only worthy goals of education.

SOCRATES' SECOND OBJECTION: MEMORY'S DESTRUCTION

*[In] modern Guatemala . . . Mayans remark that outsiders
note things down not in order to remember them, but rather
so as not to remember them.*
—NICHOLAS OSTLER

*If men learn this, it will implant forgetfulness in
their souls; they will cease to exercise memory because they rely
on that which is written, calling things to remembrance
no longer from within themselves, but by means of
external marks. What you have discovered is a recipe
not for memory, but for reminder.*
—PHAEDRUS

The unbridgeable differences that Socrates saw between spoken
and written words in their different pedagogical and philosophical uses, in their ability to depict reality, and in their capacity to

refine thought and virtue were mild in comparison with his concern for the changes literacy would bring to memory, and the individual's internalization of knowledge. Socrates well knew that literacy could greatly increase cultural memory by reducing the demands on individual memory, but he didn't want the consequences of the trade.

By committing to memory and examining huge amounts of orally transmitted material, young educated Greek citizens both preserved the extant cultural memory of their society and increased personal and societal knowledge. Unlike the judges at his trial, Socrates held this entire system in esteem not so much from a concern for preserving tradition as from the belief that only the arduous process of memorization was sufficiently rigorous to form the basis of personal knowledge that could then be refined in dialogue with a teacher. From this larger interconnected view of language, memory, and knowledge, Socrates concluded that written language was not a "recipe" for memory, but a potential agent of its destruction. Preserving the individual's memory and its role in the examination and embodiment of knowledge was more important than the indisputable advantages of writing in preserving cultural memory.

Most of us take memorization for granted as a component of education from kindergarten to graduate school. But in comparison with the Greeks, or even with our own grandparents, we are required to do little or no explicit memorization of passages. Once a year I ask my undergraduate students how many poems they know "by heart"—a curiously lovely, anachronistic phrase. Students of ten years ago knew between five and ten poems; students today know between one and three. This small sample makes me wonder anew about Socrates' seemingly archaic choices. What are the implications for the next generations, who may commit even less to memory—whether it is fewer poems or even, for some, only part of the multiplication tables? What happens to these children when the electricity goes out, the computer breaks down, or the rocket's systems malfunction? What is the difference in the brain's pathways connecting language and long-term memory for our children and the children of ancient Greece?

Certainly my children's eighty-six-year-old Jewish grand-
mother, Lotte Noam, would flummox future generations. On almost any occasion she can supply an appropriate three-stanza
poem from Rilke, a passage from Goethe, or a bawdy limerick—
to the infinite delight of her grandsons. Once, in a burst of envy, I
asked Lotte how she could ever memorize so many poems and
jokes. She answered simply, "I always wanted to have something
no one could take away if I was ever put into a concentration
camp." Lotte prompts us to pause and consider the place of memory in our lives, and what the incremental atrophying of this quality, generation by generation, ultimately means.

There is a vivid example of how Socrates reacts to this loss of
personal memory when he catches young Phaedrus using what
may be the world's first recorded crib sheet to recite a speech by
Lysias. To aid his memory, Phaedrus wrote the speech down and
tucked it inside his tunic. Suspecting what his pupil had done,
Socrates launched into a diatribe on the nature of written words
and their pitiful inability to aid instruction. He began by likening
writing to beautiful paintings that only appear lifelike, "But if
you question them, they maintain a majestic silence. It is the same
with written words; they seem to talk to you as though they were
intelligent, but if you ask them anything about what they say,
from a desire to be instructed, they go on telling you just the same
thing over and over again forever."

One can only sympathize with Phaedrus, who was not the
only brunt of Socrates' ire. In *Protagoras,* Socrates mercilessly at-
tacked those who think "like papyrus rolls, being able neither to
answer your questions nor to ask themselves."

SOCRATES' THIRD OBJECTION:
THE LOSS OF CONTROL OVER LANGUAGE

Ultimately, Socrates did not fear reading. He feared superfluity of
knowledge and its corollary—superficial understanding. Reading
by the untutored represented an irreversible, invisible loss of control over knowledge. As Socrates put it, "Once a thing is put in
writing, the composition, whatever it may be, drifts all over the
place, getting into the hands not only of those who understand it,

but equally of those who have no business with it; it doesn't know how to address the right people, and not address the wrong. And when it is ill treated and unfairly abused it always needs its parents to come to its help, being unable to defend or help itself."

Underneath his ever-present humor and seasoned irony lies a profound fear that literacy without the guidance of a teacher or of a society permits dangerous access to knowledge. Reading presented Socrates with a new version of Pandora's box: once written language was released there could be no accounting for what would be written, who would read it, or how readers might interpret it.

Questions about access to knowledge run throughout human history—from the fruit of the tree of knowledge to Google. Socrates' concerns become greatly amplified by our present capacity for everyone with a computer to learn very, very quickly about virtually anything, anywhere, anytime at an "unguided" computer screen. Does this combination of immediacy, seemingly limitless information, and virtual reality pose the most powerful threat so far to the kind of knowledge and virtue valued by Socrates, Plato, and Aristotle? Will modern curiosity be sated by the flood of pat, often superficial information on a screen, or will it lead to a desire for more in-depth knowledge? Can a deep examination of words, thoughts, reality, and virtue flourish in learning characterized by continuous partial attention and multitasking? Can the essence of a word, a thing, or a concept retain importance when so much learning occurs in thirty-second segments on a moving screen? Will children inured by ever more realistic images of the world around them have a less practiced imagination? Is the likelihood of assuming we understand the truth or reality of a thing even greater if we see it visually depicted in a photograph, film, or video or on "reality TV"? How would Socrates respond to a filmed version of a Socratic dialogue, to his entry in Wikipedia, or to a screen clip on YouTube?

Socrates' perspective on the pursuit of information in our culture haunts me every day as I watch my two sons use the Internet to finish a homework assignment, and then tell me they "know all about it." As I observe them, I feel an unsettling kinship with Socrates' futile battles so long ago. I cannot help thinking that we have lost as much control as Socrates feared 2,500 years ago over what, how, and how deeply the next generation learns. The profound gains are equally obvious, beginning with Plato's preservation of Socrates' objections.

In the last analysis, Socrates lost the fight against the spread of literacy both because he could not yet see the full capacities of written language and because there would be no turning back from these new forms of communication and knowledge. Socrates could no more prevent the spread of reading than we can prevent the adoption of increasingly sophisticated technologies. Our shared human quest for knowledge ensures that this is as it must be. But it is important to consider Socrates' protests as we grapple with the brain and its dynamic relationship to reading. Socrates' enemy never really was the writing down of words, as Plato realized. Rather, Socrates fought against failures to examine the protean capacities of our language and to use them "with all our intelligence."

In this Socrates was not alone, even then. Across the world, in India of the fifth century BCE, scholars of Sanskrit also decried written language, valuing oral language as the truest vehicle for intellectual and spiritual growth. These scholars mistrusted and condemned any dependence on written text that could short-circuit their lifework—the analysis of language.

As we turn next to the development of language and reading in the "youngest members of our species," I hope that Socrates' concerns will inform our own personal Greek chorus, urging us to examine how the life of words in young children and the pursuit of knowledge and virtue can come alive for this new generation—and for generations beyond.

PART **II**

HOW *the* BRAIN LEARNS
to READ OVER TIME

*Among the many worlds which man did not receive
as a gift of nature, but which he created with his own spirit, the
world of books is the greatest. Every child, scrawling his first
letters on his slate and attempting to read for the first time, in so
doing, enters an artificial and most complicated world;
to know the laws and rules of this world completely and to
practice them perfectly, no single human life is long enough.
Without words, without writing, and without books there
would be no history, there could be no concept of humanity.*
—HERMANN HESSE

Chapter 4

THE BEGINNINGS OF READING
DEVELOPMENT, OR NOT

*When the first baby laughed for the first time, the laugh broke
into a thousand pieces, and that was the beginning of the fairies.*
—J. M. BARRIE, *Peter Pan*

*It seems to me that, beginning with the age of two, every child
becomes for a short period of time a linguistic genius. Later,
beginning with the age of five to six, this talent begins to fade.
There is no trace left in the eight-year-old of this creativity with
words, since the need for it has passed.*
—KORNEI CHUKOVSKY

I MAGINE THE FOLLOWING SCENE. A SMALL CHILD
sits in rapt attention on the lap of a beloved adult, listening
to words that move like water, words that tell of fairies, dragons,
and giants in faraway places never before imagined. The young
child's brain prepares to read far earlier than one might ever suspect, and makes use of almost all the raw material of early childhood, every perception, concept, and word. It does so by learning
how to use all the important structures that will make up the
brain's universal reading system. Along the way, the child incorporates many of the insights into written language that our spe-

cies learned, breakthrough by breakthrough, during more than 2,000 years of history. It all begins under the crook of an arm in the comfort of a loved one's lap.

Decade after decade of research shows that the amount of time a child spends listening to parents and other loved ones read is a good predictor of the level of reading attained years later. Why? Consider more carefully the scene we just described: a very young child is sitting, looking at colorful pictures, listening to ancient tales and new stories, learning gradually that the lines on the page make letters, letters make words, words make stories, and stories can be read over and over again. This early scene contains most of the precursors crucial to the child's development of reading.

How a child first learns to read is a tale of either magic and fairies or missed chances and unnecessary loss. These two scenarios tell of two very different childhoods—the first, in which almost everything we hope for occurs; and the second, in which few tales are told and little language is learned, and the child falls farther and farther behind before reading can even begin.

The First Story

Working with premature infants highlights the importance of touch in their development. A similar principle applies to the ideal development of reading. As soon as an infant can sit on a caregiver's lap, the child can learn to associate the act of reading with a sense of being loved. In a zany, endearing scene in the movie *Three Men and a Baby*, Tom Selleck reads the results of dog races to his infant charge. Everyone yells at him for corrupting the baby, but he is right on target. You can read an eight-month-old racing results, stock prices, or Dostoyevsky, although an illustrated version would be even better.

Why has Margaret Wise Brown's *Goodnight Moon* captured the imagination of millions of children who beg their parents to read it night after night? Is it the use of pictures of beloved items in a room—the night lamp, the mitten, the bowl of mush, the rocking chair: things that belong to the world of childhood? Is it the sense of discovery as children learn to find a tiny mouse that hides in a different place on every page? Is it the reader's voice, which seems to get softer and softer until the book's last page? All these reasons and more provide an ideal beginning for a long process that some researchers call emergent or early literacy. The association between hearing written language and feeling loved provides the best foundation for this long process, and no cognitive scientist or educational researcher could have designed a better one.

SERIOUS WORDPLAY

The next step in the process involves a growing understanding of pictures, as the child becomes able to recognize the visual images illustrating a few books that will soon be dog-eared. Underlying this development is a visual system that is fully functional by six months, an attention system that has a long road ahead to maturation, and a conceptual system that grows by leaps and bounds each day. As the ability to pay attention increases with each passing month, so too does the infant's knowledge of familiar visual images, and his or her curiosity about novel ones.

As children's perceptual and attention abilities grow, they engage with the most important precursor for reading, early language development, and with it the pivotal insight that things like ponies and dogs have names. It is an experience in every child's life similar to what Helen Keller must have experienced when she first realized that water—her tactile experience of it—had a name, a label that she could communicate through sign language to everyone. It is what the ancient writers of the Rig Veda recognized: "The Wise established Name-giving, the first principle of language."

It can be difficult for adults to suspend their own views of the everyday world in order to realize that very young children don't "know" each thing in this world has a name. Very gradually, children learn to label the salient parts of their world, usually beginning with the people who care for them. But the realization that everything has its own name typically comes at around eighteen

months and is one of the insufficiently noted eureka events in the
first two years of life. The special quality of this insight is based
on the brain's ability to connect two or more systems to make
something new. Underlying a child's epiphany is the young brain's
ability to connect and integrate information from several systems:
vision, cognition, and language. Contemporary child linguists
such as Jean Berko Gleason emphasize that every time a child
learns what it is to name—whether what is named is a beloved
human, a kitten, or Babar—a major cognitive change also has
begun connecting the developing oral language system to developing conceptual systems.

With the emergence of naming, the content of books begins to
play a larger role, for now children can direct the choice of what
is read. There are important developmental dynamics here: the
more children are spoken to, the more they will understand oral
language. The more children are read to, the more they understand all the language around them, and the more developed their
vocabulary becomes.

This intertwining of oral language, cognition, and written
language makes early childhood one of the richest times for language growth. The cognitive scientist Susan Carey of Harvard
studies how children learn new words, something she humorously
calls "zap mapping." She finds that most children between two
and five years old are learning on average between two and four
new words every day, and thousands of words over these early
years. These are the raw material of what the Russian scholar
Kornei Chukovsky called the child's "linguistic genius."

Linguistic genius comes from various elements of oral language that will all later fold into the development of written language. Phonological development, a child's evolving ability to
hear, discriminate, segment, and manipulate the phonemes in
words, helps pave the way for the critical insight that words are
made up of sounds—for example, that "cat" has three distinct
sounds (/k/-/a/-/t/). Semantic development, a child's growth in vocabulary, contributes an ever-increasing understanding of the
meanings of words, which fuels the engine of all language growth.
Syntactic development, a child's growth in acquiring and using

the grammatical relationships within language, paves the way to understanding the growing complexity of sentences in the language of books. For example, it enables the child to understand how word order affects meaning: as in "The cat bit the mouse," which differs from "The mouse bit the cat." Morphological development, a child's acquisition and use of the smallest units of meaning (like the plural "s" in "cats," and the past tense "ed" in "walked"), contributes to an understanding of the kinds of words and grammatical uses of those words found in sentences and stories. Finally, pragmatic development, a child's ability to perceive and use the social-cultural "rules" of language in its natural contexts, provides the basis for understanding how words can be used in the countless situations described in books.

Each aspect of oral language development makes an essential contribution to the child's evolving understanding of words and their multiple uses in speech and in written text.

LAUGHTER, TEARS, AND FRIENDS

None of these linguistic abilities, however, develops in a vacuum. All are based on underlying changes in the developing brain, the child's growing conceptual knowledge, and the particular contributions made by each child's developing emotions and understanding of other people. All these factors are either nurtured or neglected by the child's environment. To bring this idea to life, let's first place a three-and-a-half-year-old girl, with all her "linguistic genius," on the lap of a person who often reads to her. This child already understands that particular pictures go with particular stories and that stories convey feelings that go with the words—feelings that range from happiness to fear and sadness. Through stories and books she is beginning to learn a repertoire of emotions. Stories and books are a safe place for her to begin to try these emotions on for herself, and are therefore a potentially powerful contributor to her development. At work here is a reciprocal relationship between emotional development and reading. Young children learn to experience new feelings through exposure to reading, which, in turn, prepares them to understand more complex emotions.

This period of childhood provides the foundation for one of the most important social, emotional, and cognitive skills a human being can learn: the ability to take on someone else's perspective. Learning about the feelings of others is not simple for three- to five-year-olds. The best-known child psychologist of the twentieth century, Jean Piaget, described children this age as egocentric, in the sense that they are constrained by their level of intellectual development to a view of the world as revolving around themselves. It is their gradually evolving ability to think about others' thinking—not their moral character—that prevents them from being able to know what another person is feeling.

An example can be found in *Frog and Toad*, a series of books by Arnold Lobel. In one story Frog is very sick, and Toad comes to his rescue without a second thought, motivated only by empathy. Toad feeds Frog every day and cares for him, until at last Frog can get out of bed and play again. This little story offers a quiet, profound model of what it means to understand what someone else feels, and how this can become the basis for helping one another.

Books about another animal species—the hippopotamus—convey similar insights about empathy. In James Marshall's famous series *George and Martha*, two lovable hippos are best friends. In each story they teach us something about being a good and understanding friend. In one memorable story, George stumbles and breaks off one of his two front teeth, which are very important for a hippo. After getting a new gold replacement, he worriedly shows it to Martha, who knows just what to say to her friend. " 'George!' she exclaimed. 'You look so handsome and distinguished with your new tooth!' " And of course George is then happy.

These stories exemplify thoughts and feelings experienced by many young children as they listen to stories and books. We may never fly in a hot-air balloon, win a race with a hare, or dance with a prince until the stroke of midnight, but through stories in books we can learn what it feels like. In this process we step outside ourselves for ever-lengthening moments and begin to understand the "other," which Marcel Proust wrote lies at the heart of communication through written language.

What the Language of Books Teaches Us All

Around the time we begin to recognize feelings that both connect us to other people and distinguish the boundaries between us, another insight comes that is more overtly cognitive: a book is full of long and short words that stay the same every time it is read, just like the pictures. This gradual intellectual discovery is part of a larger, more tacit discovery, that books have a language all their own.

"The language of books" is a concept rarely articulated by children and, to be sure, little considered by most of us. In fact, several somewhat unusual and important conceptual and linguistic features accompany this language and contribute immeasurably to cognitive development. First, and most obviously, the special vocabulary in books doesn't appear in spoken language. Think back to tales that delighted you, stories that began like this:

> Once, long ago, in a dark, lonely place where the sunlight was never seen, there lived an elfin creature with hollow cheeks and waxen complexion; for no light ever touched this skin. Across the valley, in a place where the sun played on every flower, lived a maiden with cheeks like rose petals, and hair like golden silk.

No one, or at least no one I know, ever speaks this way. Phrases like "Once, long ago," and words like "elfin" aren't part of typical discourse. However, they are an integral part of the language of books and give children clues that help them predict what type a story is and what might happen. Indeed, by kindergarten, words from books will be one of the major sources of the 10,000-word repertoire of many an average five-year-old.

A large portion of these thousands of words are morphological variations of already known root words. For example, the child who learns the root word "sail" can quickly understand and acquire all its derived forms: *sails, sailed, sailing, sailboat*, and so on. But it is not only vocabulary growth that is special about the language of story and books. Equally important is the syntax or grammatical structure found here, which is largely absent from the stuff of everyday speech. "Where the sun was never to be seen"; "for no light ever touched this skin": these are constructions typically found only in print, and they require no small amount of cognitive flexibility and inference. Few children under age five hear "for" used as it is used in "for no light ever touched this skin," where it serves as a connective, a class of grammatical devices like "then" and "because" that teach causal relationships between events and concepts. Children learn this use of "for" from the context. When they do, syntactic, semantic, morphological, and pragmatic aspects of language development all become enriched.

Studies by the reading researcher Victoria Purcell-Gates underscore the more serious implications of this point. Purcell-Gates looked at two groups of five-year-old children before they could read. The two groups were similar on variables like socioeconomic status and parental educational level; but one group had been "well-read-to" in the two years before (at least five times a week), and the other, control group had not. Purcell-Gates simply asked the two groups of children to do two things: first, to tell a story about a personal event like a birthday; and second, to pretend that they were reading a storybook to a doll.

The differences were unmistakable. When the children in the "well-read-to" group told their own stories, they used not only more of the special "literary" language of books than the other children but also more sophisticated syntactic forms, longer phrases, and relative clauses. What makes this significant is that when children are able to use a variety of semantic and syntactic forms in their own language, they are also better able to understand the oral and written language of others. This linguistic and cognitive ability provides a unique foundation for many comprehension skills a few years later, when children begin to read stories of their own.

One recent study by the sociolinguist Anne Charity and her colleague Hollis Scarborough indicates the importance of knowledge of grammar for children who speak other dialects and other languages. They found that in a group of children who used

African-American Vernacular English dialect rather than Standard American English dialect, each child's knowledge of grammar predicted how well he or she would eventually learn reading.

Another feature of the language of books involves a beginning understanding of what might be called "literacy devices," such as figurative language, particularly metaphor and simile. Consider these similes from the example above: "cheeks like rose petals, and hair like golden silk." Such phrases are both linguistically lovely and cognitively demanding. Children are being asked to compare "cheeks" to "rose petals" and "hair" to "silk." In the process, they are gaining not only vocabulary skills, but also practice in the cognitively complex use of analogy. Analogical skills represent an extremely important, largely invisible aspect of intellectual development at every age.

A charming example of early analogical skills is found in *Curious George*, about a monkey whose irrepressible curiosity concerning balloons leads him to sail above the sky, where the "houses looked like toy houses and the people like dolls." The simple similes here actually aid the child in performing such sophisticated cognitive operations as size-based comparisons and depth perception. The author, Hans Rey, and his Bauhaus-trained wife Margret may have been unaware of the contributions they were making to children's cognitive and linguistic development when they first created the rascal George in the 1940s, but to this very day they continue to influence the development of millions of preschool children.

Another contribution from the language of books involves higher-level understanding by the child. Think about the phrase "Once, long ago." In a flash it transports you from your present reality and activates a special set of expectations about another world. "Once, long ago" cues every savvy preschooler that this is going to be a fairy tale. Arguably, there are only several hundred different types of stories, with many variations across cultures and times. Children eventually develop an understanding of many of these distinct types, each of which has its own typical plot, setting, era, and set of characters. This kind of cognitive information is part of what goes into "schemata," a term some psychologists use to refer to how certain ways of thinking become routinized and help a person make sense of events and remember them better. The principles here function in a self-reinforcing spiral: the more coherent the story is to the child, the more easily it is held in memory; the more easily remembered the story is, the more it will contribute to the child's emerging schemata; and the more schemata a child develops, the more coherent other stories will become and the greater the child's knowledge base for future reading will be.

The ability to predict likely scenarios aids the child's development of inferential skills (deduction or guessing based on whatever information is given). Children who have five years of experience fighting with trolls, rescuing maidens with silken tresses, and deciphering clues given by witches will have an easier task recognizing unfamiliar words (like "tresses" and "trolls") in print, and ultimately, and most important, comprehending the texts that contain them.

After considering the many ways that exposure to books helps children's development of later reading, we might assume that just reading a great many books to your child is enough preparation in the preschool reading period. Not quite. According to some researchers, being read to is only part of what prepares children for reading. Another good predictor is the seemingly humble ability to name a letter.

WHAT'S IN A LETTER'S NAME?
As children gain familiarity with the language of books, they begin to develop a more subtle awareness of the visual details of print. Many a child in many a culture can be seen "reading" a book by moving a finger, even when there is not a single line of print in sight. One aspect of print awareness begins with the discovery that printed words go in a particular direction: for example, in English and European languages from left to right; in Hebrew and Aramaic languages from right to left; and in several Asian scripts from top to bottom.

Next comes a trickier set of skills. As the particular shapes of

Figure 4-1: Two Chinese Characters

some lines become increasingly familiar, some children can identify some of the colored letters on the refrigerator door, in the bathtub, or on a piece of drawing paper. The brain's ability to recognize the visual shape of, say, a turquoise letter is no casual feat, as every ancient token-reading brain can attest to. As we have seen, it is based on an exquisitely fine-tuned visual perception system and considerable exposure to the same patterns and features in the visual world that allow us to recognize owls, spiders, arrows, and crayons.

Before they learn to recognize a letter automatically, much less label it, children have to make some of their neurons in the visual cortex "specialists" in detecting the tiny, unique set of features of each letter—exactly what the first token readers had to do. To get a sense of what a child has to learn at the level of visual analysis, look at the Chinese characters in Figure 4-1. These two Chinese logographs consist of many of the same visual features that are used in alphabetic letters: curves, arcs, and diagonal lines. Pause for a few seconds and then turn immediately to the last page of this chapter. Are the two symbols shown there exactly the same as those on this page, or are there any differences? (The answer is in the Notes section for this page.) Most adults find this a humbling exercise. It shows the sophisticated perceptual demands made on the young visual system, which must learn that each of the tiny but salient features in every letter of our alphabet conveys information, and that letters consist of orderly patterns of these features that do not change—at least not too much.

Here, an important, early set of conceptual skills—pattern

invariance—facilitates learning letters. As an infant, our young child already learned that some visual features (mother's face, and father's) don't change. They are invariant patterns. As discussed earlier in Chapter 1, humans have innate abilities that permit us to store representations of perceptual patterns in our memory and then apply them to each new learning situation. From the start, therefore, children search for invariant features when they try to learn something new. This helps them build visual representations and rules that will eventually allow them to identify any letter on a refrigerator, regardless of size, color, or font.

From another view of cognitive development, a child's first effort at naming letters is not much more than "paired-associate" learning: that is, it has all the conceptual glamour of a pigeon learning to pair some object with a label to get a pellet. Quickly enough, however, far more cognitively elegant learning of letters emerges, something akin to Susan Carey's notion of "bootstrapping" in the learning of numbers. For example, for many children counting to ten and the "alphabet song" provide a conceptual "placeholder" list. Gradually, each number and letter name in the list will be mapped onto its grapheme (written) form, accompanied by a growing understanding of what the letter or number does. The late neuropsychologist Harold Goodglass once told me that for much of his early childhood, he was convinced that "elemeno" was a long letter in the middle of the alphabet. This is an example of how children's concepts of letters change right along with their developing language, their invisible conceptual development, and their use of specialized visual areas of the brain for letter identification.

A comparison of object and letter naming in the young child reveals a rather unexpected "pre"- and "post-blueprint" of the brain's evolution before and after literacy. At a simple level, recognizing and naming objects are the processes children first use to begin connecting their underlying visual areas to areas serving language processes. Later, in a process akin to Stanislas Dehaene's notion of neuronal recycling, letter recognition and naming enlist special portions of these same circuits so that written symbols can eventually be read very quickly.

We have no brain images of the child first learning letter names, but we do have brain imaging of object and letter naming in adults. It shows that in the first milliseconds both processes share a great deal of area 37's fusiform gyrus. One hypothesis is that early letter naming in children will look quite similar to object naming in the preliterate child. As the child learns to recognize letters as discrete patterns or representations, working groups of neurons gradually become more and more specialized and require less and less area. In this sense, naming objects and then naming letters represent the first two chapters of the modern, rearranged literate brain.

The brilliant German philosopher Walter Benjamin (1892–1940) held that naming was the quintessential human activity. Although he never saw a brain scan, Benjamin could not have been more correct with regard to the early development of naming and reading. Learning to retrieve a name for an abstract, visually presented letter-symbol is an essential precursor for all the processes that come together in reading, and a powerful predictor of a child's readiness to read. The work of my research group over many years indicates that the ability to name objects when a child is very young, and then to name letters, as the child matures, provides a fundamental predictor of how efficiently the rest of the reading circuit will develop over time.

To be sure, the age when a child can name letters varies greatly among children and among cultures. In some cultures and in countries like Austria, children do not learn letters until the first grade. In the United States, some two-year-olds can name all the letters, but some five-year-olds (particularly boys) must still work hard at this. Indeed, I have heard several boys between five and seven softly sing the entire alphabet song before they are finally able to find a sought-for letter and name it.

Parents should be encouraged to help children name letters whenever they appear ready, and the same principle applies to "reading" what is called environmental print—familiar words and signs in the child's environment such as a stop sign, a box of cereal, the child's name, and the names of siblings or friends. Many prekindergarten children and most kindergarten children recognize the shapes of very familiar words such as "exit" and "milk," and often the first letters of their own names. It doesn't matter that some children may insist that "Ivory" says "soap." Gradually, each child in most literate cultures begins to acquire a repertoire of frequently seen letters and words before ever learning to write these letters. This phase of reading is like a "logographic" stage in the child's development: what the child understands, not unlike our token-reading ancestors, is the relationship between concept and written symbol.

WHEN SHOULD A YOUNG CHILD BEGIN TO READ?
As soon as children begin to name the letters of the alphabet, the question arises whether they should learn to read "early." The hope of many parents, and the sales pitches of many a commercial pre-reading program, is that reading early will give children an advantage later on in school. Twenty-six years ago, a colleague of mine at Tufts, the child psychologist David Elkind, wrote an insightful book, *The Hurried Child*, on the tendency in our society to push children to achieve. He cited the ever-earlier ages at which parents urge their children to read. Recently, David decided to make a new edition of the book because he believes the situation is significantly worse now than it was two decades ago.

Our biological timetables add to this discussion. Reading depends on the brain's ability to connect and integrate various sources of information—specifically, visual with auditory, linguistic, and conceptual areas. This integration depends on the maturation of each of the individual regions, their association areas, and the speed with which these regions can be connected and integrated. That speed, in turn, depends a great deal on the *myelination* of the neuron's axons. Nature's best conductive material, myelin, consists of a fatty sheathing wrapped around the cell's axons (Figure 4-2). The more myelin sheathes the axon, the faster the neuron can conduct its charge. The growth of myelin follows a developmental schedule that differs for each region of

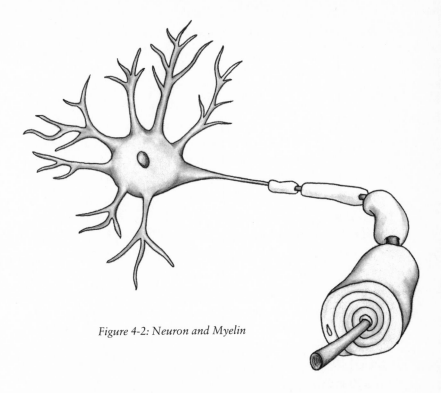

Figure 4-2: Neuron and Myelin

the brain (for instance, auditory nerves are myelinated in the sixth prenatal month; the visual nerves, six months postnatally).

Although each of the sensory and motor regions is myelinated and functions independently before a person is five years of age, the principal regions of the brain that underlie our ability to integrate visual, verbal, and auditory information rapidly—like the angular gyrus—are not fully myelinated in most humans until five years of age and after. The behavioral neurologist Norman Geschwind suggested that for most children myelination of the angular gyrus region was not sufficiently developed till school age—that is, between five and seven years. Geschwind also hypothesized that myelination in these critical cortical regions develops more slowly in some boys; this might be one reason why more boys are slower to read fluently than girls. To be sure, our own research on language finds that girls are faster than boys until around age eight on many timed naming tasks.

Geschwind's conclusions about when a child's brain is sufficiently developed to read received support from a variety of cross-linguistic findings. The British reading researcher Usha Goswami drew my attention to a fascinating cross-language study by her group. They found across three different languages that European children who were asked to begin to learn to read at age five did less well than those who began to learn at age seven. What we conclude from this research is that the many efforts to teach a child to read before four or five years of age are biologically precipitate and potentially counterproductive for many children.

In reading readiness, as in life, there are always exceptions. A memorable fictional example of a child who learns to read before age five is Scout in Harper Lee's *To Kill a Mockingbird*, who horrifies her new teacher with her precocious ability to read anything in sight.

> As I read the alphabet a faint line appeared between her eyebrows and after making me read most of the My First Reader and the stock market quotations from the Mobile Register aloud, she discovered that I was literate and looked at me with more than faint distaste. Miss Caroline told me to tell my father not to teach me anymore, it would interfere with my reading. I never deliberately learned to read. . . . Reading was something that just came to me. . . . I could not remember when the lines above Atticus's moving finger separated into words, but I had stared at them all the evenings in my memory— anything Atticus happened to be reading when I crawled into his lap every night. Until I feared I would lose it, I never loved to read. One does not love breathing.

The writer Penelope Fitzgerald gives another view on the subject. She recalls, "I began to read just after I was four. The letters on the page suddenly gave in and admitted what they stood for. They obliged me completely and all at once." For children like Scout and Penelope Fitzgerald, by all means let them read! For the rest, there are excellent biological reasons why reading comes in its own good time.

Notes from the Great Unmyelinated

Many wonderful things can happen before age five that are developmentally appropriate and facilitate both later reading and enjoyment of preschool without explicit reading instruction. Writing and listening to poetry, for example, sharpen a child's developing ability to hear (and ultimately to segment) the smallest sounds in words, the phonemes. Such first attempts to write reflect a sequence in a child's growing knowledge about the connection between oral and written language. First, letters are written (or drawn) in imitation. To be sure, there is often more scribbled "art" than concept here. Next, letters begin to show off children's evolving concept of print, particularly the letters in their own names. Gradually, other letters capture how children think words are spelled, with many a letter name used, let us say, ingeniously.

In a book called *Gnys at Work: A Child Learns to Write and Read*, Glenda Bissex provides a picturesque example of the period when children use the names of letters to spell words. At one point when Bissex was preoccupied (probably with writing her book), her five-year-old son slipped her this note: "RUDF." These letters translate straightforwardly: "Are you deaf?" Bissex's son, like countless children his age, had begun to have two insights: first, that writing can command an adult's sometimes transitory attention; and second, the complex notion that letters correspond to the sounds inside words. What he missed was that the sound a letter represents and the name of the letter are not equivalent. The letter "r" doesn't represent "are"; rather, it represents the sound of the English phoneme /r/, pronounced "ruh." This correspondence between written letters and oral sounds is a subtle and difficult concept. Often parents, and even teachers who are untrained in the linguistic basis of reading, forget the complexity involved. Indeed it is a concept largely missing in most of the earlier primers used to teach reading to children.

Preschool children of ages four and five may not figure out these subtler insights, but they do begin to learn symbolic representation at a new level. They learn that printed words represent spoken words; that spoken words are made up sounds; and, very importantly, that letters convey those sounds. In many children, this realization leads to an outpouring of writing that is highly unconventional in terms of English spelling rules, but in actuality is extremely rule-governed. This writing, called "invented spelling" by Carol Chomsky and Charles Read, is quite decipherable if you remember Glenda Bissex's son. However, the principle is trickier than it looks. For example, try to decipher "YN." This spelling has been found in children's writing for at least two words: "wine" and "win." In both instances the children used the letter name Y to convey the "wuh" sound (which English represents with "w"). In writing "wine" they used the complete letter name Y, but in "win" they used the complete letter name N to convey "in": two perfectly reasonable sets of possible spelling rules.

Another unusual feature of invented spelling in early writing is that the sounds often don't conform to the accepted spelling, because English pronunciation is fearsomely various and affected by many factors, including regional dialect. For example, in Boston, where I live, the medial "t" in many a word (such as "little") is spelled with a "d" by children ("LDL"); and a South Boston child and his or her Brahmin counterpart will both take about a year more than children elsewhere in the country to render "cart" with an "r." Many a Bostonian child will, however, like the late President John Kennedy, generously bestow said "r" on the end of "AMREKR."

One of the more intriguing questions about children's first writing is whether or not they can read it. In fact, most children are hard-pressed to read back what they have written, but oh, do they want to! This motivation, coupled with learning the individual sounds in words that goes into "invented spelling," makes children's early writing an extremely useful precursor of learning to read, and a wonderful complement to the actual reading process.

PHONEME AWARENESS AND THE WISE MOTHER GOOSE

Young children do not perceive the same sound units that we do, as we've seen with Harold Goodglass's "elemeno" and the charmingly idiosyncratic examples of children's writing. Rather, children move very gradually from an awareness of what makes up a word in a sentence to syllables inside a word (e.g., "sun-ny"), until finally each individual phoneme inside a word can be segmented (e.g., "s," "u," "n"). A child's awareness of the discrete sounds and phonemes in a word is both a critical component and an outgrowth of learning to write and learning to read. As we saw in the achievement of the Greeks, the meta-awareness of individual speech sounds didn't just magically appear in the history of writing; nor does it appear magically in the child. When asked by the reading expert Marilyn Adams what the "first sound" in "cat" was, one child promptly replied, "Meow"!

One of the singular accomplishments of the creators of the Greek alphabet was this dimension of awareness of speech sounds. It is one of the most powerful contributions of the alphabet, and also is one of the two best predictors of later reading achievement, the other being rapid naming. Invented spellings like RUDF and all types of a child's early writing give clues as to when this linguistic awareness develops, and they also promote its growth.

In addition to writing, there are other, equally entertaining ways to help children develop an awareness of phonemes. Mother Goose is one. Tucked inside "Hickory, dickory dock, a mouse ran up the clock" and other rhymes can be found a host of potential aids to sound awareness—alliteration, assonance, rhyme, repetition. Alliterative and rhyming sounds teach the young ear that words can sound similar because they share a first or last sound. If you listen to young children tell their first jokes, the whimsical appeal of rhyme will immediately strike you. Like Winnie-the-Pooh, children love to repeat a "matching pair" of sounds over and over and over ("Funny bunny, you're a funny bunny, honey!"), just because the rhyme catches their fancy.

Equally important, the child who has begun to discriminate paired sounds has also begun to segment the internal parts of words into smaller components. Children four and five years old are learning to discern the onset or first sounds of a word ("S" in "Sam") and the rime ("am" in "Sam"). This is the beginning of the long important process of being able to hear each individual phoneme in a word, which facilitates learning to read.

A well-known, highly inventive experiment in England by several famous researchers illustrates the importance of these princi-

ples. Lynne Bradley and Peter Bryant investigated four groups of
preschool children who were similar in every way except that at age
four two groups received a training program emphasizing alliterative and rhyme sounds. These children listened to groups of words
that had either the same initial (alliterative) sound or the same medial vowel (rhyme) in the final sound. They learned simply to group
the words according to shared sounds. In addition, children in one
of the two trained groups were shown the letter that matched in the
sound categorization tasks. Several years later Bradley and Bryant
tested all the children. Rather astonishingly, those children who received the simple rhyme training displayed much more developed
phoneme awareness and, most important, learned to read more
easily. Further, children who had both the rhyme training and the
visual matched letter condition did best of all. Fostering Chukovsky's "linguistic genius" in the young child happens in many
ways, and the poetry of the nursery rhyme is one of them.

But what is happening under the surface in the development
of children to produce this unlikely finding? At the most basic
level, the children learn first to perceive words more analytically
in the most effortless way possible—through attending to alliteration and rhyme and learning to categorize sounds on that basis.
They then connect these sounds to a matched letter or visual representation. Together, the skills used in listening to the melody,
rhythm, and cadence of the rhymes tucked within the verses of
Mother Goose facilitate the child's "phoneme awareness skills."
Extensive research on the development of this phonological aspect of language indicates that systematic play with rhymes, first
sounds, and last sounds in wordplay, jokes, and songs significantly
contributes to a child's readiness to learn to read. Teaching a child
to enjoy poetry and music is serious child's play.

The Scottish language researcher Katie Overy and two members of our lab, Catherine Moritz and Sasha Yampolsky, are cur-
rently finding that specific emphases within musical training
itself—such as rhythm pattern production—may also enhance
phoneme awareness and other precursors to reading development. If this proves to be true, they hope to create early interventions based on rhythm, melody, and rhyme.

KINDERGARTEN: WHERE PRECURSORS
COME TOGETHER

When children are five to six years of age, all the precursors of reading come together in the world of kindergarten. No pre-learned concept, letter, or word is wasted by good teachers. Earlier learning becomes the stuff of the more formal introduction to the world of written language. Although teachers have been nurturing most precursors for many years, only in recent years have systematic tools become widely available to promote the development of phoneme awareness skills. These seemingly simple methods help children learn several difficult linguistic concepts: (1) the "Moses insight" (as in Thomas Mann's story) that there can be a one-to-one correspondence between a sound and a symbol; (2) the more difficult concepts that each letter has both a letter name and a sound or group of sounds that it represents: and the converse that each sound is represented by a letter or sometimes several letters; and (3) the understanding that words can be segmented into syllables and sounds.

The reading researcher Louisa Cook Moats articulates the importance of infusing these basic linguistic principles into the teaching of reading and into the development of early reading skills like blending. Children often have a very hard time figuring out how to blend sounds together to make words like "cat" and "sat." Knowing the linguistic principle that a "continuant" phoneme like "s" can be held on to as long as it takes for the child to add the rime (such as "at") makes teaching the concept of blending a lot easier for child and teacher. Thus, if you want to teach blending, "sat" and "rat" make early blending more manageable than the proverbial "cat."

The Second Story

Thus far, what flows into the acquisition of reading takes place in a very special world where mother rabbits and loving hippos illuminate words and feelings, dragons convey concepts and syntax, and nursery rhymes and scrawled approximations of letters teach an awareness of sounds and print, as well as a dawning awareness of their relationship. Reading in such a world is the sum of five years spent developing highly complex cognitive, linguistic, perceptual, social, and affective abilities, all of which flourish best in rich environmental interactions.

What then of children who come from homes where no one hears Mother Goose, where no one is encouraged to read signs, write scriggly letters, or play with books of any kind? What of children in this country from other cultures, who hear many stories, but in Spanish, Russian, or Vietnamese? What of children who don't appear to learn or respond to language in the same way as others? More and more groups of children like these begin to fill our classrooms, each with different needs. What happens to them as they enter kindergarten has serious consequences for the rest of their lives—for them, and for all of us.

THE WAR ON "WORD POVERTY"

Unbeknownst to them or their families, children who grow up in environments with few or no literacy experiences are already playing catch-up when they enter kindergarten and the primary grades. It is not simply a matter of the number of words unheard and unlearned. When words are not heard, concepts are not learned. When syntactic forms are never encountered, there is less knowledge about the relationship of events in a story. When story forms are never known, there is less ability to infer and to predict. When cultural traditions and the feelings of others are never experienced, there is less understanding of what other people feel.

As mentioned earlier, a chilling finding in a study of one California community by Todd Risley and Betty Hart exposes a bleak reality with serious implications: by five years of age, some children from impoverished-language environments have heard 32 million fewer words spoken to them than the average middle-class child. What Louisa Cook Moats calls "word poverty" extends well beyond what the child hears. In another study, which looked at how many words children produce at age three, children from impoverished environments used less than half of the number of words already spoken by their more advantaged peers.

Yet another study concerns books in the home—any kind of books. In a survey of three communities in Los Angeles, there were startling differences in how many books were available to children. In the most underprivileged community, no children's books were found in the homes; in the low-income to middle-income community there were, on average, three books; and in the affluent community there were around 200 books. My carefully interwoven tale of toads, words, and syntax goes out the window when such statistics appear. The sheer unavailability of books will have a crushing effect on the word knowledge and world knowledge that should be learned in these early years.

The Canadian psychologist Andrew Biemiller studies the consequences of lower vocabulary levels in young children. He finds that children who come to kindergarten in the bottom twenty-fifth percentile of vocabulary generally remain behind the other children in both vocabulary and reading comprehension. By grade 6 approximately three full grades separate them from their average peers in both vocabulary and reading comprehension; they are even more dramatically behind children whose vocabulary in kindergarten was at or above the seventy-fifth percentile. In other words, the interrelatedness of vocabulary development and later reading comprehension makes the slow growth of vocabulary in these early years far more ominous than it appears when viewed as one unfortunate phenomenon. Nothing about language development has isolated effects on children.

Many factors that children "bring to the table" in kindergarten can't be changed. Language development is not one of them. The average household offers ample opportunities to give a child everything necessary for the normal development of language. In a broad study of early development of literacy skills, the educator Catherine Snow of Harvard and her colleagues found that in addition to literacy materials, one of the major contributors to later reading was simply the amount of time for "talk around dinner." The importance of simply being talked to, read to, and listened to is what much of early language development is about, but the reality in many families (some economically disadvantaged, some not) means that too little time will be given to even these three basic elements before a child reaches the age of five.

As the policy maker Peggy McCardle notes over and over, with relatively small, concerted efforts, the preschool years can become rich with possibilities for language development, rather than a "war zone." All professionals who deal with children can help to ensure that parents understand the contribution they can make to their child's potential and that every child can attend a good-quality preschool. For example, a series of vaccinations, a few talks to new parents about "dinner talk," and a series of free developmentally appropriate books should be the norm for every "well visit" in the first five years of life for every child who will attend American schools. Social workers and service providers in home-visiting programs such as "Healthy Start" can provide similar packages and training in these areas as well. A level playing field for all children before they enter kindergarten should not be that difficult to achieve.

EFFECTS OF EAR INFECTIONS ON EARLY LANGUAGE DEVELOPMENT

One pervasive impediment to that level playing field involves middle ear infections in young children, the single most common complaint in pediatric practices across the country. Consider what is really happening to a young child who is learning those two to four new words every day, when he or she has an undiagnosed or untreated ear infection. One day the child hears the new word "pur"; the second (or tenth) day he or she hears "pill"; another time he or she hears "purple." Thanks to an ear infection, the child receives inconsistent acoustic information, leading to three different sound representations for the word "purple." Cognitive confusion aside, these children will take longer to gain new vocabulary words, and, depending on when and how many infections occur, they may not develop a complete, high-quality repertoire of the phoneme representations that each language possesses. Untreated infections affect vocabulary development and phonological awareness, two of the most important precursors of reading.

But the problems don't stop there. If two major precursors of reading—vocabulary and phonological awareness—are affected, so is reading. One of my students in a large longitudinal research project had parents fill out a questionnaire on ear infections in the preschool years, and then got every pediatric history she could on each child. The results indicated that children with frequent untreated ear infections were significantly more likely to have reading problems later.

One of the striking insights we gained from this study concerns not the predictable outcome, but the sheer number of parents who made some comment like, "But all my children had 'runny ears' half the time." In other words, many well-meaning parents never understood that ear infections had more serious consequences than transient discomfort. Untreated "runny ears" represent an invisible impediment to the development of oral and written language, and everyone who works with children needs to know this. As with impoverished literacy environments, if relatively small but concerted efforts are made, ear infections do not need to be an obstacle for our children.

THE POSSIBLE EFFECTS OF BILINGUAL ENVIRONMENTS ON LEARNING TO READ

A much more difficult issue involves the effects of having to learn English at the same moment you enter the school door. Learning two or more languages is an extraordinary, complicated cognitive investment for children, that represents a growing reality for huge numbers of students. Some up-front costs, such as transfer errors and substitutions from one language to the next, are less important than the advantages, *if* (a very important "if") the child learns each language well. The plasticity of the young brain enables young children to attain—with less effort than at any other time—proficiency in more than one language. After puberty, students bring certain advantages to learning a language, but the younger child's brain is superior in several important ways when it comes to learning to speak accent-free languages.

Examining the many issues swirling around bilingualism and learning is dizzying, but three principles dominate. First, English-language learners who know a concept or word in their first language learn to use it more easily in English, their second, "school" language. In other words, language enrichment at home provides an essential cognitive and linguistic foundation for all learning, and it does not need to be in the school language to be of help to the child. Children who have an impoverished environment in their home language, on the other hand, have no cognitive or linguistic foundation for either the first or the second, school language.

The second principle is similar to the first. Little is more important to learning to read English than the quality of language development in English. Thousands of children enter school with varying degrees of knowledge of English. Systematic efforts to instill both the "new" phonemes of the English language and the new vocabulary of school (and books) need to happen in each classroom for each learner. Connie Juel points out an essential linguistic issue easily missed by our teachers: children who come to school either new to the English language or new to the standard American English dialect spoken in schools do not know the very phonemes they are expected to sound out (or induce) in reading. For five years, they "learned to ignore them and listen largely to their own."

The third principle concerns the age when children become bilingual: the earlier the better for oral and written language development. The neuroscientist Laura-Ann Petitto of Dartmouth and her colleagues found that early bilingual exposure (before age three) had a positive effect, with language and reading comparable to those of monolinguals. Further, in their imaging studies of adults who had been early bilinguals, Petitto's group found that the subjects' brains processed both languages in overlapping regions, like the brains of monolinguals. By contrast, bilingual adults who had been exposed later to a second language showed a different, more bilateral pattern of brain activation.

As a cognitive neuroscientist, I think that having a bilingual brain is a very good thing. Among other things, Petitto's work demonstrates how an early exposed bilingual brain appears to have certain cognitive advantages over a monolingual brain in terms of linguistic flexibility and multitasking. As an educator working in many communities where English is not spoken in most homes, however, I obsess over the complex and sometimes contentious issues involved in learning two languages, including children's self-esteem, their membership in a cultural community, their sense of perceived competence, and the cumulative effects of all this on reading. I know that we must help all our children learn the language of school, so that they can achieve their potential in this English-speaking culture, beginning with becoming readers. For some children, nurtured on an "ideal lap" in Spanish, Japanese, or Russian, learning to read in English is a more modest challenge, and listening to English storybooks helps them connect familiar words and concepts in their first language to their second. For those who had no such lap, the process of entering school and of learning a second language at the same time can have overwhelming cognitive-social-cultural effects. They are all our children, and we must be ready for each of them, beginning with a communal commitment to teaching each child and armed with knowledge about how reading in any language develops over time.

Reading never just happens. Not a word, a concept, or a social routine is wasted in the 2,000 days that prepare the very young brain to use all the developing parts that go into reading acquisition. It is all there from the start—or not—with consequences for the rest of children's reading development, and for the rest of their lives.

Chapter 5

THE "NATURAL HISTORY" OF READING DEVELOPMENT: CONNECTING THE PARTS OF THE YOUNG READING BRAIN

No one ever told us we had to study our lives,
make of our lives a study, as if learning natural history
or music, that we should begin
with the simple exercises first
and slowly go on trying
the hard ones, practicing till strength
and accuracy became one with the daring
to leap into transcendence . . .
— ADRIENNE RICH, "TRANSCENDENTAL ETUDE"

In a sense it is as if the child has recapitulated history—from the early fumblings with the discovery of alphabetic writing to the equal, if not greater, intellectual feat of discovering that the spoken word is made up of a finite number of sounds.
— JEANNE CHALL

THAT PROUST'S EXTRAORDINARY NOVEL *Remembrance of Things Past* was evoked by the taste of a madeleine is one of the almost-mythical anecdotes about literature in the twentieth century. Whether or not the narrator's sensory memory was simply a creation of Proust's considerable imagination, it could have happened. The human brain stores and retrieves memories in a variety of ways, including through each of the senses.

In beginning this chapter on learning to read, I wanted to find my own madeleine: that is, one thing that would release my first memory of true reading. I couldn't do it. I could not remember that first moment of knowing I could read, but some of my other memories—of a tiny, two-room school with eight grades and two teachers—evoke many pieces of what the language expert Anthony Bashir calls the "natural history" of the reading life. The natural history of reading begins with simple exercises, practice, and accuracy, and ends, if one is lucky, with the tools and the capacity to "leap into transcendence." All this happened for me in a little town called Eldorado.

Learning to Read in Eldorado, Florence, Philadelphia, and Antigua

When you learn to read you will be born again . . . and you
will never be quite so alone again.
—RUMER GODDEN

In books I have traveled, not only to other worlds,
but into my own. I learned who I was and who I wanted to be,
what I might aspire to, and what I might dare to dream about
my world and myself. But I felt that I, too, existed much
of the time in a different dimension from everyone else I knew.
There was waking, and there was sleeping. And then there were
books, a kind of parallel universe in which anything might
happen and frequently did, a universe in which I might be a
newcomer but was never really a stranger. My real,
true world. My perfect island.
—ANNA QUINDLEN

It was my father's wish that I be sent to school. It was an
unusual request; girls did not attend school. . . . What could an

education do for someone like me? I can only say what I did not
have; I can only measure it against what I did have and find
misery in the difference. And yet, and yet . . . it was for this
reason that I came to see for the first time what lay beyond
the path that led away from my house.
— JAMAICA KINCAID

The historian Iris Origo, marchesa of Val d'Orcia, used to quote
Rumer Godden to describe her experiences learning to read in a
Florentine villa at the beginning of the twentieth century. Anna
Quindlen wrote a perfect description of learning to read in mid-
twentieth-century Philadelphia. Jamaica Kincaid captured, in her
Autobiography of My Mother, what it meant for a girl to read in
her childhood world of Antigua in the Caribbean. Indeed, her early
prodigious facility in learning to read convinced her teacher that
Jamaica was "possessed." Despite the differences in time, place,
and cultural context among these women writers, one theme unites
each of them to every new book lover. This theme also ran through
my experiences learning to read in Eldorado, Illinois: the discovery
within books of the existence of parallel universes—Origo's never
being "quite so alone again," Quindlen's "perfect island," and
Kincaid's "what lay beyond the path that led from my home."

"Orthographic irony" is as good a term as any to describe the
origin of my hometown's name. In the mid-1800s Mr. Elder and
Mr. Reeder hired a painter from "the city" to come to the little
town they had founded—Elderreeder, in southern Illinois—to
paint a sign welcoming everyone who might ride by. Being an educated man, the painter discreetly corrected what he assumed was
a spelling error of the townfolk, and made a sign welcoming all to
"Eldorado." Perhaps the sign was too handsome to change, perhaps there wasn't enough money to buy another, or perhaps the
name appealed to some previously unarticulated dream of the
townspeople, but in any case the name stuck, and I was raised in
Eldorado, Illinois, a century later.

Two schools prepared the children of Eldorado. My tiny paro-
chial school, St. Mary's, looked like something in a nineteenth-
century woodblock: dark red brick with two large rooms, each

with four grades and four rows. First-graders sat in the leftmost row near the window, and with every passing year the children moved one row closer to the exit door.

Somewhere by that window during the middle of first grade, I began to read more than I talked, which was a great deal. At first I learned everything the kids in the second row were doing, and then everything the kids in the third row were doing. I don't remember when I finished reading all the work of the fourth-grade children, but it was while I sat in the second row. Within that setting—a classroom filled with forty youngsters—having me as a student would have tested the patience of anyone except a saint. But by almost any criteria, the teachers at this little school— Sister Rose Margaret, Sister Salesia, and later Sister Ignatius— were saints.

During my time in the second row, something remarkable happened. My teacher spoke to my parents, Frank and Mary Wolf, and suddenly books started appearing at the back of the room. Shelves that had been half empty began to fill like magic with whole series of books: fairy tales, the All-About Books on science, stories of heroes, and, to be sure, biographies of the saints. By the end of fourth grade, when my brother Joe sat in the third row, my sister Karen in the first row, and my brother Greg waited in the wings, I had read every one and more.

In the process, I changed. No matter how small I looked to the world, I daily entered the company of literal and figurative giants. Paul Bunyan, Tom Sawyer, Rumpelstiltskin, and Teresa of Avila seemed as real to me as my next-door neighbors on Walnut Street. I began to dwell in two parallel worlds, and in one of them I never felt different, or alone. This experience served me in good stead, particularly later in my life. In those years when I sat surprisingly quietly in that tiny classroom, I was newly crowned, newly wed, or newly canonized every other day.

My other vivid memory of those days centers on Sister Salesia, trying her utmost to teach the children who couldn't seem to learn to read. I watched her listening patiently to these children's torturous attempts during the school day, and then all over again after school, one child at a time.

My best friend, Jim, was one of the children who stayed late. When Sister Salesia bent over Jim, he suddenly lost any resemblance to the boy I knew—the leader of the pack, the boy who had an answer for everything, the mid-twentieth-century equivalent of Tom Sawyer and Huck Finn wrapped up in one. The other Jim looked like a pale version of himself, haltingly coming up with the letter sounds Sister Salesia asked for. It turned my world topsy-turvy to see this indomitable boy so unsure of himself. For at least a year they worked quietly and determinedly after school ended. Sister Salesia told his family that some very bright children like Jim needed extra help learning to read.

This was all that was ever said, but even then I realized two things. First, I saw how determined Sister Salesia was, and how tenaciously she and Jim's mother held on to his potential, even when Jim was ready to quit. I thought to myself that they were doing something special. Second, by the time Jim moved to the third row, I noticed that my old friend was back, as cocky, audacious, and irrepressible as ever. Then I knew that Sister Salesia and his mother were doing something miraculous.

. . . .

Learning to read *is* an almost miraculous story filled with many developmental processes that come together to give the child entry into the teeming underlife of a word usable by the child. Socrates and the ancient Indian scholars feared that reading words, rather than hearing and speaking them, would prevent our ability to know their many layers of meaning, sound, function, and possibility. In fact, early reading exposes—during the moment of acquisition—how many of the multiple, older structures contribute to each layer as they come together to form the brain's new circuitry for reading. Studying the development of early reading, therefore, allows us to peek into the underpinnings of our species' accomplishment, beginning with the interrelated processes that prepared the child in the first five years and that expand in different, predictable ways over the rest of the development of reading.

Phonological development—how a child gradually learns to hear, segment, and understand the small units of sounds that make up words—critically affects the child's ability to grasp and learn the rules of letter sounds at the heart of decoding.

Orthographic development—how the child learns that his or her writing system represents oral language—gives a critical foundation for all that follows. The child must learn the visual aspects of print—such as the features of letters, the common letter patterns, and "sight" words in English—and also how to spell all these new words.

Semantic and pragmatic development—how children learn more and more about the meanings of words from the language and culture around them—heightens and quickens children's ability to recognize a word they are laboriously decoding and to comprehend it with an ever quicker "aha."

Syntactic development—how children learn the grammatical forms and structures of sentences—enables them to make sense of the ways words are used to construct sentences, paragraphs, and stories. It also teaches them how events relate to each other in a text.

Morphological development, perhaps the least studied of the systems, prepares the child to learn the conventions surrounding how words are formed from smaller, meaningful roots and units of meaning (i.e., morphemes). The child who learns that "unpacked" is made of three discoverable parts—un•pack•ed—can read it and recognize it faster and better.

Together, all these developments quicken the early recognition of a word's parts, foster more facile decoding and spelling, and enhance the child's understanding of known and unknown words. The more a child is exposed to written words, the greater his or her implicit and explicit understanding of all language. In this the child is like the Sumerians, in contrast to Socrates' fears.

The reading scholar Jeanne Chall of Harvard taught that reading acquisition moves through a fairly orderly set of steps from pre-reader to expert reader, which we can study, "as if learning natural history or music." Indeed, I like to think of the interwoven relationships among the components of reading as like music: what one ultimately hears is the sum of many players, each largely indistinguishable from the rest, all contributing to the whole. Early reading is the one time in our lives when each contributing player is more discernible, enabling those of us who have long forgotten to remember what goes into every word we read.

How Reading Develops

There, perched on a cot, I pretended to read.
My eyes followed the black signs without skipping a single one,
and I told myself a story aloud, being careful to utter all the
syllables. I was taken by surprise—or saw to it that I was—a
great fuss was made, and the family decided that it was time to
teach me the alphabet. I was as zealous as a catechumen. I went
so far as to give myself private lessons. I would climb up on my
cot with Hector Malot's No Family, *which I knew by heart, and,*
half reciting, half deciphering, I went through every
page of it, one after the other. When the last page was
turned, I knew how to read. I was wild with joy.
— JEAN-PAUL SARTRE

In his memoir *The Words*, Jean-Paul Sartre recounts his first recollection of reading, and the sheer "wild joy" that accompanied this experience. However filtered through the lens of memory, Sartre's account is similar to the experience of countless children who also half memorize, half decipher a favorite book until suddenly (or so it seems to them) reading is theirs. The reality is that Sartre steadily accumulated multiple, partial sources of knowledge until "suddenly" a threshold was crossed, and he deciphered the secret language of print. The rest of this chapter chronicles the gradual, dynamic changes in us as readers between Sartre's moment of exuberant code cracking and the imperceptible move into becoming fully autonomous expert readers.

To structure this account, I introduce in the present chapter and Chapter 6 five types of readers: (1) emerging pre-reader, (2) novice reader, (3) decoding reader, (4) fluent comprehending reader, and (5) expert reader. Each type represents dynamic changes in reading development that we move through unknowingly. Not all children, however, progress in the same way. Referring to the many differences in how children learn, a well-known pediatrician, Mel Levine, writes about "all kinds of minds." Similarly, there are "all kinds of readers," some of whom follow a different sequence, with stops and starts in reading development different from the ones I describe here. Their stories will be told later.

Emerging Pre-Reader

Twice in your life you know you are approved of by everyone—
When you learn to walk and when you learn to read.
—PENELOPE FITZGERALD

As described in Chapter 4, the emerging pre-reader, sits on "beloved laps," samples and learns from the full range of multiple sounds, words, concepts, images, stories, exposure to print, literacy materials, and just plain talk during the first five years of life. The major insight in this period is that reading never just happens to anyone. Emerging reading arises out of years of perceptions, increasing conceptual and social development, and cumulative exposures to oral and written language.

Novice Reader

I can see them standing politely on the wide pages
that I was still learning to turn,
Jane in a blue jumper, Dick with his crayon-brown hair,
playing with a ball or exploring the cosmos
of the backyard, unaware they are the first characters,
the boy and girl who begin fiction.
BILLY COLLINS, FROM "FIRST READER"

Few more heartwarming or exhilarating moments exist than watching children learn that they can actually read, that they can decode the words on a page, and that the words tell a story. Not long ago I sat on the floor beside a child named Amelia, as shy as a forest creature. Not yet reading, she rarely spoke and never volunteered to read out loud for visitors like me. But that day something happened. Amelia stared a long time, as she always did, at the letters in the short sentence "The cat sat on the mat." She looked like the proverbial frozen deer. Then, very slowly, but perfectly, she began to articulate the words. She lifted her eyes to mine and her eyebrows began to rise. She then moved to the next short sentence and the next, each time looking to me for confirmation. By the end of the story, she smiled from ear to ear, and she didn't look at me to check. She had begun to read, and she knew it. Amelia had few books in any language in her home, and a long road remained ahead of her, but she had begun to read.

Whatever her precursors, whatever her literacy environment, whatever method of instruction was being used by her teacher, the tasks for Amelia, as for every novice reader, begin with learning to decode print and to understand the meaning of what has been decoded. To get there, every child must figure out the alphabetic principle that took our ancestors thousands of years to discover, with many partial discoveries along the way.

Similarly, in every domain of learning—from riding a bike to understanding the concept of death—children develop along a continuum of knowledge, moving from a partial concept to an established concept. In their beginning efforts, novice readers only partially understand the concepts underlying the alphabetic principle. I love to repeat what reading specialist Merryl Pischa asked her young charges in Cambridge, Massachusetts, every year: "Why is it that the hardest thing children are ever asked to do is the *first* thing they're asked to do!?"

By and large, most children come to reading (whether in kindergarten or in the first grade) with a notion that the words on the page mean something. Most of them have watched their parents, day-care providers, and teachers read books. Many, however, have nothing like an established concept that the words in books are made of the sounds of our language, that letters convey these sounds, and that each letter conveys a particular sound or two.

The major discovery for a novice reader is Amelia's increasingly consolidated concept that the letters connect to the sounds of the language. This is the essence of the alphabetic principle, and the foundation for the rest of Amelia's reading development. Learning all the grapheme-phoneme correspondence rules in decoding comes next for her, and this involves one part discovery and many parts hard work. Aiding both are three code-cracking capacities: the phonological, orthographic, and semantic areas of language learning.

WHEN "CAT" HAS THREE SOUNDS, NONE OF WHICH IS "MEOW": PHONOLOGICAL DEVELOPMENT

The daily, halting discoveries that happen in learning to decode the individual letters in a word propel the child's deepening awareness of phonemes, one of several important aspects of phonological development. Slowly, the child begins to hear the large and small units of sound in the speech stream, like the words in a phrase (kitty + cat), the syllables in a word (kit + ty), and the phonemes in words and syllables (/k/ + /a/ + /t/). All this, in turn, furthers reading acquisition.

Novice readers can hear and segment the larger units. Gradually, they learn to hear and manipulate the smaller phonemes in syllables and words, and this ability is one of the best predictors of a child's success in learning to read. The researcher Connie Juel at Stanford found that a child's phoneme awareness in these earliest periods was critical for learning to decode in first and second grade. Not being able to decode well in grade 1 predicted 88 percent of the poor readers in grade 4. Teachers help children become more aware of the phonemes within words through a sort of armamentarium of opportunities—such as nursery rhymes that enhance the child's ability to hear and segment the rhymes and alliterative structure of words, and little "instant games" in which clapping, writing, and dancing beat out the sounds in words.

Phonological or sound blending involves the child's larger ability to synthesize—literally, to blend—individual sounds to form larger units such as syllables and words (blending s + a + t = sat). Just like phoneme awareness skills, blending develops steadily over time through practice, and through more and more reading.

Approaches to blending have proliferated over the years. One of the most interesting involves a technique used by the educator George O. Cureton in Harlem. He assigned each child the sound of a letter and then lined the children up to "act out" the ways sounds blend to form words. Picture the following scene. The first child hisses the easily held sibilant /sss/, and then gently pushes into a second child, who belts out an open-throated /a/, as long as possible. The second child then thrusts himself or herself gently into the next child, who pronounces the less easily held "stop" consonant /t/. The first round of shoves probably involves chaos, but then, as the instructor directs the action faster and more gently, s-a-t becomes sat.

Children learn more easily when there are two main emphases: on the initial sound of a syllable, called the onset; and on the final vowel + consonant pattern of a syllable, called the rime (the "at" of "cat"). Depending on the instruction, children learn the onset (c), then add the rime (at), and then blend the two to make a word. Later, more difficult and varied onsets get added to the rime: ch + at = chat; fl + at = flat. This approach may be a bit more civilized than Cureton's, but it has the same goal: getting the child to integrate units of sound smoothly. Blending seems relatively simple, but it impedes reading acquisition for many children, particularly those with reading disabilities.

A useful method for helping novice readers with phoneme awareness and blending involves "phonological recoding." This may seem to be just a pretentious term for reading aloud, but "reading aloud" would be too simple a term for what is really a two-part dynamic process. Reading aloud underscores for children the relationship between their oral language and their written one. It provides novice readers with their own form of self-teaching, the "sine qua non of reading acquisition."

Extending the work of an eminent New Zealander, the educator Marie Clay, two reading experts in Boston—Irene Fountas and

Gay Su Pinnell—have long taught that reading out loud also exposes for the teacher and any listener the strategies and common errors typical for a particular child. Reading aloud helps uncover what the child knows or doesn't know about words. I will never forget how we discovered that Timmy, a typical first-grade novice reader, was regularly misreading the middle letters in words. At the outset of one standard primer story about a little house, Timmy read "horse" for "house," and then proceeded to "read" an entire story, which he invented on the spot, about a horse. Completely unrelated to the dullish text about the house, Timmy's wonderful creation helped us understand the source of many of his errors.

Andrew Biemiller studied typical errors made by children of Timmy's age and found that young novice readers tend to move through three short, fairly predictable steps. First, they make errors that are semantically and syntactically appropriate, but that bear no phonological or orthographic resemblance to the real word ("daddy" for "father"). Once they learn some rules of grapheme-phoneme correspondence, their errors show orthographic similarities to the missed word, but little semantic appropriateness (Timmy's "horse" for "house"). At the end of their time as novice readers, children make errors that show both orthographic and semantic appropriateness ("bat" for "ball"). These children are on the cusp of moving to more fluent decoding and begin to integrate the varied sources of knowledge they have about words. Very important, Biemiller found that the children who succeed most in reading never get arrested in any of these early steps, but move quickly through them.

"Who Said Yachts Are Tough?": Why Orthographic Development Prepares Kids to Read This Title

English possesses a delightfully puritanical tradition of writing one of our well-known scatological words like this: sh_t. Everyone knows that the dash stands for the missing letter "i," and this "letter stand-in" straddles the line between good taste and precise spelling. The dash also demonstrates how arbitrary all visual symbols are, and how necessary an accepted system is for depicting each of the sounds of our language. Orthographic develop-ment consists of learning the entirety of these visual conventions for depicting a particular language, with its repertoire of com-mon letter patterns and of seemingly irregular usages. Most im-portant, it involves the transformation of these visual patterns of letters and frequent letter combinations into representations that can become automatic.

Children learn orthographic conventions one step at a time. From their experience on the laps and at the sides of older read-ers, pre-readers learn that in English, words are read from left to right on a line and that letters are read from left to right in a word. The next insights involve cognitive rather than spatial discoveries: for example, pattern invariance. Many children have to learn that an "A" by any other font is still an "A." Similarly, some children must learn that uppercase and lowercase forms can represent the same letter. But the real task involves learning the unique ways that English conveys its sounds in varied but English-specific let-ter patterns. Look at a word in two languages that share many roots: the English "shout" and the German *schreien*. Although there are commonalities between the English "sh" and the Ger-man "schr," these two letter patterns are largely unique ortho-graphic representations in each language, like "ois" is in French, and "lla" and "ña" in Spanish.

Novice readers absorb all the most common letter patterns in their own language, and also many of the most frequently written words that don't necessarily follow the phonological rules, such as "have," "who," and all the words in "Who said yachts are tough?" Although a large majority of the most frequent common words can be decoded with the child's phonological knowledge, a few very important common words cannot. These latter, irregu-larly spelled words, often called "sight words," need to become orthographic representations of their own. Luckily, there are fewer irregularly spelled words than is commonly thought, if you are aware of English rules, and most irregularly spelled words, like "yacht," qualify as only partially irregular.

However one labels it, orthographic development for novice

readers requires multiple exposures to print—practice, by any other name. At the University of Washington, the neuroscientist and educator Virginia Berninger and her group document how the young brain needs all these exposures to form orthographic representations of the most common visual chunks, so that simple letter patterns like "ant" become "chant" and "enchantment" within an eyeblink. To be sure, this demands more than just the eye, but the ability of the visual system to unpack consonant clusters like the "ch" in "chant," and also morpheme units like "en" and "ment" in "enchantment," increases the speed of reading tremendously. Explicit learning of common vowel patterns, morpheme units, and varied spelling patterns in English (e.g., the prickly clusters of consonants that precede many a word) aids the work of the visual system.

That said, English vowels must be some of the most overworked symbols in any language on earth. How could anyone invent a writing system that forces five vowels (plus y on occasion) into double and triple duty to make up more than a dozen vowel sounds? Mark Twain's ire about English letter patterns is experienced every day, in every English-speaking classroom. The anonymous poem below captures Twain's biliousness and the feelings of thousands of novice English readers. Learning all the vowel pairs and vowel + r and vowel + w combinations can solve part of the challenge; but learning about both the varied semantic meanings and the common morphemes in words speeds up the reading of many a novice reader for many a multi-syllab-ic word.

> *I take it you already know*
> *Of touch and bough and cough and dough?*
> *Others may stumble, but not you*
> *On hiccough, thorough, slough, and through?*
> *Well done! And now you wish, perhaps,*
> *To learn of less familiar traps?*
>
> *Beware of heard, a dreadful word*
> *That looks like beard and sounds like bird.*

And dead; it's said like bed, not bead;
For goodness sake, don't call it deed!
Watch out for meat and great and threat,
(They rhyme with suite and straight and debt).
A moth is not a moth in mother.
Nor both in bother, broth in brother.

And here is not a match for there,
And dear and fear for bear and pear,
And then there's dose and rose and lose—
Just look them up—and goose and choose,
And cork and work and card and ward,
And font and front and word and sword.
And do and go, then thwart and cart.
Come, come, I've hardly made a start.

A dreadful language? Why, man alive,
I'd learned to talk it when I was five.
And yet to read it, the more I tried,
I hadn't learned it at fifty-five.

Discovering That a "Bug" Can Spy! Semantic Development in Novice Readers

Earlier, I used intriguing research by the cognitive scientist David Swinney to show that every word read activates many possible meanings, even when we are oblivious of this fact. Laying down all those various meanings is part of the beauty of childhood—or, in its absence, a great waste. For some children, knowledge of a word's meaning pushes their halting decoding into the real thing. As we saw with Amelia, in her earliest stages of beginning to decode, every word loomed as a challenge. For Amelia and for thousands of code-cracking novice readers like her, semantic development plays much more of a role than many advocates of phonics recognize, but far less of a role than advocates of whole language assume. Three related principles in semantic development transcend all pedagogical differences.

Knowing the Meaning Enhances the Reading.
If the meaning of the child's awkwardly decoded word is readily available, his or her utterance has a better chance of being recognized as a word and also remembered and stored. As Connie Juel stresses, one of the biggest errors in reading instruction is the assumption that after Amelia, for instance, finally decodes a word, she knows what she is reading. Vocabulary contributes to the ease and speed of decoding. Here's an experiment to illustrate the same principle for adults. Try to read the following terms aloud: "periventricular nodular heterotopia"; "pedagogy"; "fiduciary"; "micron spectroscopy." How fast you read each of these words depends not only on your "decoding" ability but also on your background knowledge. If these words were not in your vocabulary, chances are good you used the individual morphemes (e.g., peri + ventricle + ar) to guess at the meanings and also to improve your pronunciation. We adults also read with far more ease and efficiency words that we know something about.

Reading Propels Word Knowledge.
Vocabulary background represents a special "leg up" for many children. As the clinician and linguist Rebecca Kennedy asserts, vocabulary is one aspect of oral language "that comes for free" in learning to read. I sometimes ask my students to explain the term for a particular syndrome: for example, the word "agoraphobia." If they hesitate, I give them a sentence with the word in context: "Dr. Spock's patient with agoraphobia refused to come to the group meeting in the wide-open lecture hall." Invariably, the sentence provides them with just enough context to move the word "agoraphobia" to their next level of knowledge. Our ability to use context is fostered by reading. As novice readers' texts become more demanding, their partial concepts, combined with their "derivation" and "contextual" abilities, push many words into the established category, thereby increasing their repertoire of known words. When one realizes that children have to learn about 88,700 written words during their school years, and that at least 9,000 of these words need to be learned by the end of grade 3, the huge importance of a child's development of vocabulary becomes crystal-clear.

Multiple Meanings Enhance Comprehension.
This principle harkens back to the two stories of early reading development. Louisa Cook Moats calculates the sobering difference between linguistically advantaged children and disadvantaged children entering first grade at about 15,000 words. How can our linguistically disadvantaged children ever catch up? Explicit instruction in vocabulary in the classroom addresses some of the problem, but novice readers need to learn much more than the surface meaning of a word, even for their simple stories. They also need to be knowledgeable and flexible regarding a word's multiple uses and functions in different contexts. They need to know and to feel comfortable about bugs that crawl, pester, drive, and spy on people.

My research coordinator, Stephanie Gottwald, recounts that many of the struggling readers we work with in our intervention look horrified at the idea that an English word can have more than one meaning. When taught about words like "bug," "jam," "ram," and "bat," their first reaction is, "You've got to be kidding!" Young beginning decoders comprehend more when they understand that printed words—just like words spoken in jokes and puns—can have multiple meanings. A concept of multiplicity in words poises the novice reader to try to infer and gain more meaning from what is read, and this is the stuff of the next level of reading. But first, let's look at what the beginning decoding brain does when it reads little words such as "bat," "rat," or "bug."

THE BRAIN OF THE NOVICE READER
Cat Stoodley's drawing depicts what happens when beginning novice readers look at a word, regardless of how well they decode and understand it. As in adults' universal reading system, three large areas appear when a young child reads. The principal job for the young reading brain is to connect these parts. In a child's brain, unlike an adult's, the first large activation area covers far more territory in the occipital lobes (i.e., visual and visual association areas), as well as in an evolutionarily important area deep inside the occipital lobes and adjacent to the temporal lobe: the fusiform gyrus. Very important, there is also a great deal more

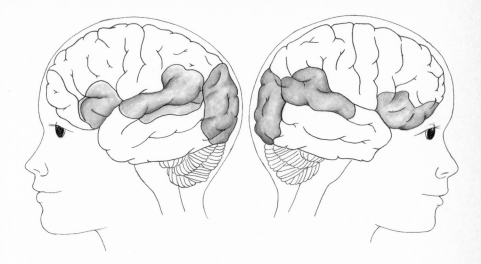

Figure 5-1: Reading Brain at Earliest Stages

activity in both hemispheres. This may seem counterintuitive at first, but think about what goes into becoming skilled at anything. In the beginning, learning any skill needs a great deal of cognitive and motoric processing and underlying neuronal territory. Gradually, as the skill becomes highly practiced, there is less cognitive expenditure, and the neuronal pathways also become streamlined and efficient. This is slow-motion development toward specialization and automaticity in the brain.

The second large distribution area, which is also bihemispheric, appears somewhat more active in the left hemisphere and includes a variety of regions in the temporal and parietal lobes. Recently, neuroscientists at Washington University found that children use more of several specific regions than adults do, particularly the angular gyrus and supramarginal gyrus, which are two important structures for integrating phonological processes with visual, orthographic, and semantic processes. Parts of an essential language comprehension region in the temporal lobe called Wernicke's area were also highly activated in children.

What is most interesting is that these first two large areas in the universal reading system are used much more by children, except under one set of conditions. Adults use these areas more than children do when words become so difficult that we revert back to childhood strategies—as some of you might have just experienced when you tried to read "periventricular nodular heterotopia."

Portions of the frontal lobes, particularly the important left-hemisphere speech area called Broca's area, make up the child's third major region. This makes sense, given the role of the frontal areas in various executive processes, such as memory, and in various language processes, such as phonology and semantics. To be sure, adult readers activate some frontal areas more, areas that are involved in these more complex comprehension and executive processes. Other areas in lower layers of the brain play active roles in both children and adults. Examples are the cerebellum and the multipurpose thalamus—one of the brain's switchboards, which links all five of its layers. "Cerebellum" means "little brain," and the cerebellum is just that, as it contributes to the timing and precision of many motoric and linguistic skills necessary for reading.

In sum, the first picture of the young novice reader's brain should impress any viewer. From the very start, the brain's capacity for making new connections shows itself here, as regions originally designed for other functions—particularly vision, motor, and multiple aspects of language—learn to interact with increasing speed. By the time a child is seven or eight, the beginning decoding brain illustrates both how much the young brain accomplishes and how far we have evolved from the first token reader. These three major distribution regions will be the foundation across all phases of reading for basic decoding, even though increasing fluency—the hallmark of the next readers—adds an interesting caveat to the unfolding portrait of the reading brain.

Decoding Reader

If you listen to children in the decoding reader phase, you will "hear" the difference. Gone are the painful, if exciting, pronunciations that characterized Amelia's reading. In their place comes the sound of a smoother, more confident reader on the verge of becoming fluent.

My favorite decoding reader was a Vietnamese boy named Van. I first met him at the Malden Summer School, where members of our research center teach children who need intense work in literacy skills. In four weeks, under the perspicacious eye of his teacher Phyllis Schiffler, Van was transformed from a grade 2 novice reader, whose teachers wanted to hold him back a grade, to a boy who performed at or even above grade level on every reading test. The audibly labored quality of Van's reading at the start of summer school vanished, replaced not only by increased attention to prosodic elements, but also by an increased amount of time for understanding what he read. Van read expressively and comprehended almost everything. Van's reading had changed from the staccato hesitations of a child who had just learned to decode to the almost smooth performance of a mid-third-grader, a perfect semi-fluent decoding reader. After only a little persuasion, and with the indisputable evidence of his reading tests, the principal and teachers at Van's school readily agreed to let him go on to grade 3. We were ecstatic and so was his family.

But there was a strange twist to Van's tale. The next summer, Van returned to us in summer school. Once again, the gifted teachers who directed the summer school program, Katharine Donnelly Adams and Terry Joffe Benaryeh, were informed that Van was in danger of being held back. Again they assigned him to Phyllis Schiffler, and, astoundingly, he read fluently for her! The directors and I were mystified. Finally, Phyllis Schiffler pulled Van aside and asked him why his teacher in grade 3 had thought he did so poorly, when in fact he read so beautifully. Shyly, he asked, "How else can I get to come to summer school?" None of us had ever encountered a faked reading disability: Van was the first.

FROM "BE" TO "BEHEADED": CONSOLIDATING PHONOLOGICAL AND ORTHOGRAPHIC DEVELOPMENT FOR DECODING READERS

In this phase of semi-fluency, readers need to add at least 3,000 words to what they can decode, making the thirty-seven common letter patterns learned earlier are no longer enough. To do this, they need to be exposed to the next level of common letter patterns and to learn the pesky variations of the vowel-based rimes and vowel pairs. In the following excerpt, consider the variety of fairly common words with the vowel pair "ea" and its huge range of possible pronunciations:

> There once was a beautiful bear who sat on a seat
> near to breaking and read by the hearth about how the
> earth was created. She smiled beatifically, full of ideas for
> the realm of her winter dreams.

This array of possible pronunciations of "ea" explains why some educators throw their hands up with regard to English orthography and want children to learn everything in context, however ineffectually. However, if you think about any letter pattern within the context of the whole word, regular rules can often be found. For example, when "ea" is followed by "r," there are only two typical possibilities (e.g., "dear" and "bear"). When it is followed by m, n, p, or t, it will usually have only one. It is essential during this phase for the semi-fluent decoding readers to acquire a good repertoire of the letter-pattern and vowel-pair "sight chunks" that make up words beyond the primer level. In addition, they learn to "see" these chunks automatically. "Sight-words" add important elements to the achievements of novice readers. "Sight-chunks" propel semi-fluency in the decoding reader. The faster a child can see that "beheaded" is be + head + ed, the more likely it is that more fluent word identification will allow the integration of this awful word—which, by the way, appears more often than you might suspect in the next step of reading.

WHAT'S IN A WORD? THE SEMANTIC, SYNTACTIC, AND MORPHOLOGICAL DEVELOPMENT OF A DECODING READER, OR NOT

The extraordinary importance of children's knowledge about "what's in a word" is that it moves them from basic decoding to fluent reading. The Tale of Two Childhoods can be rewritten right here, or it can be calcified for a lifetime. The reading scholar Keith Stanovich used a biblical reference, "Matthew effects," to describe the constructive or destructive relationship between reading development and vocabulary, where the rich get richer and the poor poorer. For word-rich children, old words become automatic, and new words come flying in, both from the child's sheer exposure to them and from his or her figuring out how to derive the meanings and functions of new words from new contexts. These readers are poised for fluent reading.

For children who are word-poor, their impoverished semantic and syntactic development has consequences for their oral and their written language. If vocabulary doesn't develop, partially known words don't become known, and new grammatical constructions are not learned. Fluent word recognition is significantly propelled by both vocabulary and grammatical knowledge. The increasingly sophisticated materials that decoding readers are beginning to master are too difficult if the words and their uses are seldom or never encountered by the children. For the word-poor child, reality actually worsens because of the usually undiscussed fact that precious little explicit vocabulary instruction goes on in most classrooms, as Isabelle Beck and her colleagues recently described. Children who know "what's in a word" read years ahead of those who do not.

With each step forward in reading and spelling, children tacitly learn a great deal about what's inside a word—that is, the stems, roots, prefixes, and suffixes that make up the morphemes of our language. Children already know about very common bound morphemes such as "s" and "ed," because these frequently appear attached to another word (thus "moons" has two morphemes: moon and a "bound" morpheme, s). Decoding readers become exposed to many types of morphemes such as prefixes ("un," "pre") and suffixes ("er," "ing"); and when they learn to read these as "sight chunks," their reading and their understanding speed up. For example, children implicitly learn that some morphemes change the grammatical function of a word: for instance, the addition of "er" changes an active verb like "sing" into a noun like "singer." And they begin to see that many words share common orthographically displayed roots that convey related meanings despite different pronunciations (e.g., sign, signer, signed, signing, signature).

But children too rarely receive explicit instruction in this second half of what makes English a "morphophonemic" writing system. As Marcia Henry, an expert on morphology, teaches, words like "sign" and "signature" provide perfect ways to illustrate to children the morphophonemic nature of the English writing system and the very good reasons for seemingly unnatural silent letters like "g" in "sign" and "c" in "muscle." Morphological knowledge is a wonderful dimension of the child's uncovering of "what's in a word," and one of the least exploited aids to fluent comprehension.

The "Dangerous Moment": Approaching Fluent Comprehension

Perhaps it is only in childhood that books have any deep influence on our lives. . . . I remember distinctly the suddenness with which a key turned in a lock and I found I could read—not just the sentences in a reading book with the syllables coupled like railway carriages, but a real book. It was paper-covered with the picture of a boy, bound and gagged, dangling at the end of a rope inside a well with the water rising above his waist—an adventure of Dixon Brett, detective. All a long summer holiday I kept my secret, as I believed: I did not want anybody to know that I could read. I suppose I half consciously realized even then that this was the dangerous moment.
— Graham Greene

I have written a great deal about fluency. With my colleague Tami Katzir, from Haifa University, I have written a new developmental definition for it. What I want to say here is very simple. Fluency is not a matter of speed; it is a matter of being able to utilize all the special knowledge a child has about a word—its letters, letter patterns, meanings, grammatical functions, roots, and endings—fast enough to have time to think and comprehend. Everything about a word contributes to how fast it can be read.

The point of becoming fluent, therefore, is to read—really read—and understand. The end of the decoding reader phase leads directly to the portal of Greene's "dangerous moment" and to the "parallel universe" described by Jamaica Kincaid and Anna Quindlen. At this point children can decode Greene's "syllables coupled like railway carriages" so quickly that they can now infer what the hero's situation involves, predict what the villain will do, feel what the heroine suffers, and think about what they themselves are reading.

To be sure, decoding readers are skittish, young, and just beginning to learn how to use their expanding knowledge of language and their growing powers of inference to figure out a text. The neuroscientist Laurie Cutting of Johns Hopkins explains some nonlinguistic skills that contribute to the development of reading comprehension in these children: for example, how well they can enlist key executive functions such as working memory and comprehension skills such as inference and analogy. Working memory provides children with a kind of temporary space for holding information about letters and words, just long enough so that the brain can connect it to the children's increasingly sophisticated conceptual information.

As decoding readers progress, their comprehension becomes inextricably bound to these executive processes, and to what they know about words and to fluency. They are all related. Incremental increases in fluency allow for inference making, because there is added time for inferences and insights. Fluency does not ensure better comprehension; rather, fluency gives enough extra time to the executive system to direct attention where it is most needed— to infer, to understand, to predict, or sometimes to repair discordant understanding and to interpret a meaning afresh.

For example, in *Charlotte's Web* a decoding reader must realize what Wilbur's fate would be without Charlotte's intervention. But what prepares the child to comprehend the splendidly sophisticated arachnoid reasoning behind this intervention? This phase of reading marks the time when the young child begins to learn how to predict from the delicate mix of what is said in a text and what is *not* said. It is the moment when children first learn to go "beyond the information given." It is the beginning of what will ultimately be the most important contribution of the reading brain: time to think.

Sometimes, however, a child in this phase of development also needs to know simply that he or she must read a word, sentence, or paragraph a second time to understand it correctly. Knowing when to reread a text (e.g., to revise a false interpretation or to get more information) to improve comprehension is part of what my Canadian colleague Maureen Lovett refers to as "comprehension-monitoring." Her research on children's meta-cognitive abilities—particularly their ability to think about how well they are understanding what they read in a text—emphasizes the importance at this phase of development of a child's being able to change strategies if something does not make sense, and of a teacher's powerful role in facilitating that change. By the end of this period, decoding readers think in a new way when they read.

WHAT ABOUT FEELING?

At any age, the reader must come across: *the child reader is the most eager and quick to do so; he not only lends to the story, he flings into the story the whole of his sensuous experience which from being limited is the more intense.*
—ELIZABETH BOWEN

As every teacher knows, emotional engagement is often the tipping point between leaping into the reading life or remaining in a childhood bog where reading is endured only as a means to other ends. An enormously important influence on the development of comprehension in childhood is what happens after we remember, predict, and infer: we feel, we identify, and in the process we understand more fully and can't wait to turn the page. The child who is moving from decoding well to decoding fluently often needs heartfelt encouragement from teachers, tutors, and parents to make a stab at more difficult reading material. Amelia needed me to affirm her efforts; Van needed support from Phyllis Schiffler.

But there is another aspect to the feeling dimension: children's ability to throw themselves fully into *Charlotte's Web*, or into any story, any book, "whole hog." After all the letters and decoding rules are learned, after the subterranean life of words is grasped, after the various comprehension processes are beginning to be deployed, the elicitation of feelings can bring children into a life-long, head-on love affair with reading and develop their ability to become fluent comprehending readers. This ever-fresh ability forms the basis for Adrienne Rich's "leap into transcendence" and the next, final steps in reading development, which make many of us who we are. The children who never make this leap never come to know what a little girl was feeling when she sat in the third-graders' row, newly canonized, freshly wed, and kissed for the first time by a prince in Eldorado, Illinois.

Chapter 6

THE UNENDING STORY OF READING'S DEVELOPMENT

I feel certain that if I could read my way back, analytically, through the books of my childhood, the clues to everything could be found. The child lives in the book; but just as much the book lives in the child.
—Elizabeth Bowen

A quality of attention has been given you.
—William Stafford

I like to take my own sweet time.
—Luke, age nine

ONE OF MY FAVORITE CHILDREN AMONG THE participants in our research was a boy named Luke. In a portent of things to come, he joined our intervention program in a most unusual way. Typically, children who qualify for our study are struggling readers who have been recommended by their teachers, and who have then passed a battery of strenuous tests. Not Luke. He basically recommended himself for our reading intervention. When asked why, he solemnly responded, "I have to read my arias. I can't memorize them anymore!" Luke, it turned out, sang in the Boston Children's Opera. He was a gifted singer, but he could no longer keep up with children who could read their lyrics.

Luke's teachers in school thought he read well enough, if a bit slowly, and did not recommend him for our intervention research. They were unaware of the chasm between how Luke should have been performing and his laborious, albeit accurate reading. After a series of tests, our clinically astute research associate Kathleen Biddle quietly said she'd never tested a child with a more profound problem in the time it took to name a letter and read a word. She then described a rather astonishing discrepancy between Luke's intelligence and his reading scores. After considerable efforts in our intervention program, Luke finally learned how to read with enough fluency to get through his arias and to make the transition from able decoding to fluent reading. But in the process he taught all of us how hard it can be to move from accuracy to fluency in the higher stages of reading.

Many children never make this transition, and for reasons that have little to do with Luke's form of reading disability. Recent reports from the National Reading Panel and the "nation's report cards" indicate that 30 to 40 percent of children in the fourth grade do not become fully fluent readers with adequate comprehension. This is a devastating figure, made even worse by the fact that teachers, textbook authors, and indeed the entire school system have different expectations for students from grade 4 on. This approach is encapsulated in the mantra that in the first three grades a child "learns to read," and in the next grades the child "reads to learn." After children leave the third grade, teachers expect them to have sufficiently automatic reading skills that enable them to learn more and more "on their own," from increasingly difficult text materials. I had exactly the same expectation when I taught. Through no fault of their own, most fourth-grade teachers never take a course in teaching reading to children who have not acquired fluency.

One nearly invisible issue in American education is the fate of young elementary students who read accurately (the basic goal in most reading research) but not fluently in grades 3 and 4. Unless their problems are dealt with, these students will be left in the dust. We know a lot about developmental dyslexia and intervention, but we know far less about the more ordinary problems of children who never attain fluency for various reasons that do not lend themselves to diagnosis: such as, a poor environment, a poor vocabulary, and instruction not matched to their needs. Some of these children become capable decoding readers, but they never read quite rapidly enough to comprehend what they read. Some of them, like Luke, have an undiagnosed "rate of processing" type of dyslexia that we will discuss later on. Whatever the reasons, to have close to 40 percent of our children "underachieving" reflects a horrific waste of human potential. It is a great "black hole" of American education—a netherworld of the semiliterate, into which more and more of our children slip.

Fluent, Comprehending Reader

So much of a child's life is lived for others. . . .
All the reading I did as a child, behind closed doors, sitting on
the bed while the darkness fell around me, was an act of
reclamation. This and only this I did for myself. This was
the way to make my life my own.
—LYNNE SHARON SCHWARTZ

Few more popular books can be found on bookshelves in middle schools than the *Guinness Book of World Records*. With its awe-inspiring, easily found, categorized facts, this book offers an unlikely analogy for the newly fluent reading brain. The reader at the stage of fluent comprehending reading builds up collections of knowledge and is poised to learn from every source.

Children who read from books like *Guinness* usually decode so smoothly and effortlessly that without our imaging technology we can no longer see what lies beneath. At this time teachers and parents can be lulled by fluent-sounding reading into thinking that a child understands all the words she or he is reading. Socrates railed against exactly this silent aspect of written words, which can't "speak back." For decoding does not mean comprehension. Even when a reader comprehends the facts of the content, the goal at this stage is deeper: an increased capacity to apply an understanding of the varied uses of words—irony, voice, metaphor, and point of view—to go below the surface of the text. As their reading becomes more demanding, good readers' developing knowledge of figurative language and irony helps them discover new meanings in the text that propel their understanding beyond the words themselves.

As the psychologist Ellen Winner describes in *The Point of Words*, metaphor gives "a window on children's classification skills," and irony illuminates the author's unique "attitude about the world." For example, look at the passage below from Mark Twain's *Huckleberry Finn*. Twain's uniquely ironic humor and metaphor lead many young readers to difficult and sometimes unwanted insights. In this excerpt, Huck is traveling on a raft on the great Mississippi River with his friend Jim, a runaway slave, who is being hunted down. To keep a group of men from discovering Jim's identity, Huck, in a stroke of genius, pretends that Jim has smallpox. After the men scurry away, Huck is beset with doubts:

> They went off and I got aboard the raft, feeling bad
> and low, because I knowed very well I had done wrong,
> and I see it warn't no use for me to try to learn to do
> right; a body that don't get *started* right when he's little
> ain't got no show—when the pinch comes there ain't
> nothing to back him up and keep him to his word, and so
> he gets beat. Then I thought a minute, and says to myself,
> hold on; s'pose you'd 'a' done right and give Jim up,
> would you felt better than what you do now? No, says I,
> I'd feel bad—I'd feel just the same way I do now. Well,
> then, says I, what's the use you learning to do right when
> it's troublesome to do right and ain't no trouble to do
> wrong, and the wages is just the same? I was stuck.

Huck's twisted logic and self-blame are vintage Twain. Newly fluent readers learn from Twain's irony and from his powerful images and metaphors to go below the surface of what they read to appreciate the subtext of what the author is trying to convey. For young readers who are moving from simply mastering content to discovering what lies beneath the surface of a text, the literature of fantasy and magic is ideal.

Think of the many images that Tolkien uses in *Lord of the Rings* to portray good and evil. The worlds of Middle Earth, Narnia, and Hogwarts provide fertile ground for developing skills of metaphor, inference, and analogy, because nothing is ever as it seems in these places. To figure out how to elude ring-wraiths and dragons, and how to do what is right, calls on all of one's wits. During their different journeys, Huck and Frodo learned to choose virtuous action, however powerfully they are challenged. And so do the young readers who accompany them all the way.

The world of fantasy presents a conceptually perfect holding environment for children who are just leaving the more concrete stage of cognitive processing. One of the most powerful moments in the reading life, potentially as transformative as Socrates' dialogues, occurs as fluent comprehending readers learn to enter into the lives of imagined heroes and heroines, along the Mississippi or through a wardrobe portal.

Comprehension processes grow impressively in such places as these, where children learn to connect prior knowledge, predict dire or good consequences, draw inferences from every danger-filled corner, monitor gaps in their understanding, and interpret how each new clue, revelation, or added piece of knowledge changes what they know. To practice these skills, they learn to unpeel the layers of meaning in a word, a phrase, or a thought. That is, in this long phase of reading development, they leave the surface layers of text to explore the wondrous terrain that lies beneath it. The reading expert Richard Vacca describes this shift as a development from "fluent decoders" to "strategic readers"— "readers who know how to activate prior knowledge before, during, and after reading, to decide what's important in a text, to synthesize information, to draw inferences during and after reading, to ask questions, and to self-monitor and repair faulty comprehension."

This segment of the journey—which often lasts till young adulthood—has as many obstacles as Frodo, Harry, Jim, and Huck encountered. From the start, young middle-school readers have to learn how to think in a new way, and although many children are poised to do so, nearly as many are not.

How does such a step occur? One well-known educational psychologist, Michael Pressley, contends that the two greatest aids to fluent comprehension are explicit instruction by a child's teachers in major content areas and the child's own desire to read. Engaging in dialogue with their teachers helps students ask themselves critical questions that get to the essence of what they are reading. For example, in reciprocal teaching, a method introduced by Annemarie Palincsar and Anne Brown, teachers explicitly help students learn to question what they don't understand, summarize the content, identify key issues, clarify, and predict and infer what happens next. When successful, this variation on the Socratic dialogue provides students with a lifelong approach to extracting meaning from more and more sophisticated text.

Children's desire to read reflects their immersion in the "reading life." Comprehension emerges out of all the cognitive, linguistic, emotional, social, and instructional factors in the child's prior development, and Proust's "divine pleasure" in immersing oneself in reading pushes it forward. A memorable scene in Carlos Ruiz Zafon's *Shadow of the Wind* brings this idea to life. The young protagonist, Daniel, is introduced to his first deep experience with books, as his father takes him to find his own "personal volume" in a mysterious library:

> Welcome to the *Cemetery of Dead Books,*
> Daniel. . . . Every book, every volume . . . has a soul. The
> soul of the person who wrote it and of those who read it
> and lived and dreamed with it. Every time a book
> changes hands, every time someone runs his eyes down its
> pages, its spirit grows and strengthens.

Daniel's father articulates a magical quality that characterizes our immersion in books, whereby books come to possess a life of their own—in which readers are the invited guests for a little while, and not the other way around. Daniel's obsession with his own "lost book" dictates the rest of the plot, which shows us how readers can enter into the "life of books" so completely that they become forever changed.

Knowing what it feels like to be young, impressionable, and frightened makes the reader more capable of understanding Daniel's life; learning Daniel's responses increases the reader's knowledge of the world. By identifying with characters, young readers expand the boundaries of their lives. They learn something new and lasting from each deeply felt encounter. Who among us, if faced with the prospect of being marooned, wouldn't think what Robinson Crusoe might have done? Who among us who has read Jane Austen doesn't think about Darcy when encountering an arrogant man—and hope to discover his hidden goodness? Elizabeth Bennet, Captain Ahab, Atticus Finch, Mona in the Promised Land, Celie and Nettie, Harry "Rabbit" Angstrom, and Jayber Crow: our ability to identify with these characters contributes to who we are.

Throwing ourselves into this dance with text has the potential to change us at every stage of our reading lives. But it is especially formative during this period of growing autonomy and fluent comprehension. The young person's task in this extended fourth phase of reading development is to learn to use reading for life—both inside the classroom, with its growing number of content areas, and outside school, where the reading life becomes a safe environment for exploring the wildly changing thoughts and feelings of youth.

The Fluent, Feeling Brain

The fluent reading brain has a cortical journey of its own to make. Not only does it expand its ability to decode and understand; it feels more than ever before. As David Rose, a prominent translator of theoretical neuroscience into applied educational technology, puts it, the three major jobs of the reading brain are recognizing patterns, planning strategy, and feeling. Any image of the fluent, comprehending reader shows this clearly through the

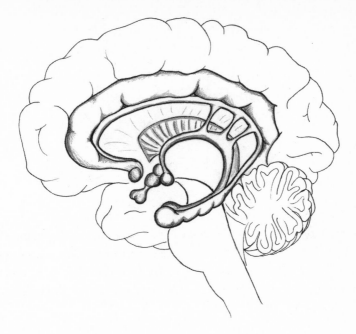

Figure 6-1: Limbic System

growing activation of the limbic system—the seat of our emotional life—and its connections to cognition. This system, located immediately below the topmost cortical layer of the brain (Figure 6-1), underlies our ability to feel pleasure, disgust, horror, and elation in response to what we read, and to understand what Frodo, Huck, and Anna Karenina experience. As David Rose reminds us, the limbic region also helps us to prioritize and give value to whatever we read. On the basis of this affective contribution, our attention and comprehension processes become either stirred or inert.

As we saw in the reading of younger children, the more effort something takes, the more the brain is activated, and usually in more expanded areas. As you will recall, the young brain's efforts to identify letters and words was reflected in the large amount of cortical space needed in the visual areas of both hemispheres, and also in a slower, less efficient pathway from visual areas to upper temporal and lower parietal regions to the frontal regions. Depicted in Figure 6-2, this slower pathway (sometimes called the

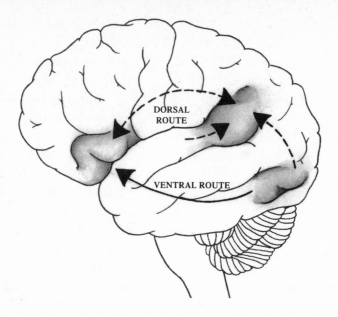

Figure 6-2: Fluent Comprehending Brain (Dorsal and Ventral Routes)

dorsal route) allows the younger child time to assemble the phonemes within a word. It also allows more "look-up" time for all
the various representations attached to words. This younger
reader, therefore, expends a great deal of time decoding.

The fluent comprehender's brain doesn't need to expend as
much effort, because its regions of specialization have learned to
represent the important visual, phonological, and semantic information and to retrieve this information at lightning speed. According to Ken Pugh, Rebecca Sandak, and neuroscientists at Yale
and Haskins Laboratory and Georgetown, as children become
more fluent, the young brain typically replaces bi-hemispheric activation with a more efficient system in the left hemisphere (sometimes called the ventral or lower route). This fluent reading
pathway begins with more concentrated, streamlined visual and
occipital-temporal regions than those used by younger children,
and then involves the lower and middle temporal regions and the
frontal regions. After we know a word very well, we no longer
need to analyze it in a labor-intensive way. Our stored letter-

pattern and word representations, particularly in the left hemisphere, activate a faster system.

Paradoxically, the developmental shift to specialized left-hemisphere activation for basic decoding processes allows more bilateral activation for meaning and comprehension processes. These shifts reflect changes in reading and human development. We are no longer mere decoders of information.

The fluent, comprehending reader's brain is on the threshold of attaining the single most essential gift of the evolved reading brain: time. With its decoding processes almost automatic, the young fluent brain learns to integrate more metaphorical, inferential, analogical, affective background and experiential knowledge with every newly won millisecond. For the first time in reading development, the brain becomes fast enough to think and feel differently. This gift of time is the physiological basis for our capacity to think "endless thoughts most wonderful." Nothing is more important in the act of reading.

The Expert Reader

And so to completely analyze what we do when we read would almost be the acme of a psychologist's achievements, for it would be to describe very many of the most intricate workings of the human mind, as well as to unravel the tangled story of the most remarkable specific performance that civilization has learned in all its history.
— SIR EDMUND HUEY

As I wrote in the preface, Sir Edmund Huey captured, in this description, how fully fluent, expert reading embodies all the cultural, biological, and intellectual transformations in the evolution of reading and all the cognitive, linguistic, and affective transformations in the reader's own "natural history." Huey's 1908 statement may well be the most eloquent description of reading ever written. Modern cognitive neuroscience reinforces what Huey suspected—

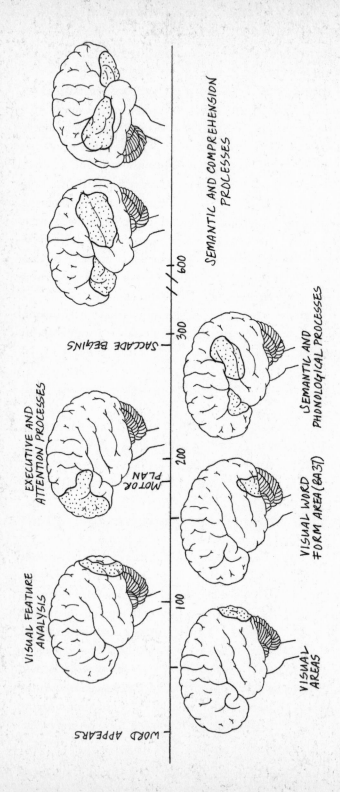

Figure 6-3: A Time Line of Reading

how vast, how complex, and how widely distributed are the brain's networks that underlie even one half second of reading.

One half second is all it takes the expert reader to read almost any word. On the basis of work by Michael Posner and various cognitive neuroscientists, I now want to describe a time line for the processes that every fully expert reader uses (Figure 6-3). Any linear conceptualization of reading (such as a time line) has to be qualified because the processes in reading are interactive. Some take place in parallel, and some activate and then reactivate when additional conceptual information needs to be integrated. For example, observe what happens when you read, "The bow on the boat was covered by a huge red bow." Most of us have to backtrack and reactivate a second reading of "bow" after we receive the added contextual information from "boat."

The time line here portrays the moment I have been waiting for: the almost instantaneous fusion of cognitive, linguistic, and affective processes; multiple brain regions; and billions of neurons that are the sum of all that goes into reading. Because it's a technical description, however, it will not be for everyone. Those who wish to may skip ahead to the end of the italicized material and read why it all leads to something magnificent in yourself and in every expert reader.

Every Word Has 500 Milliseconds of Fame

First 0 to 100 Milliseconds: Turning Expert
Attention to Letters

All reading begins with attention—in fact, several kinds of attention. When expert readers look at a word (like "bear"), the first three cognitive operations are: (1) to disengage from whatever else we're doing; (2) to move our attention to the new focus (pulling ourselves to the text); and (3) to spotlight the new letter and word. This is the orienting network of attention, and imaging research shows that each of these three operations involves a different region of the brain (Figure 6-4). To disengage attention involves areas in the back of the parietal lobe; to move our attention involves parts of the midbrain responsible for eye movements

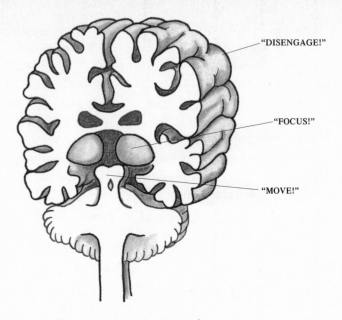

"DISENGAGE!"

"FOCUS!"

"MOVE!"

Figure 6-4: Attention Networks

(called the superior colliculi); and to spotlight something involves part of our internal switchboard known as the thalamus, which coordinates information from all five layers of the brain.

The other network of attention that is extremely important to all phases of reading is the better-known executive attention network, which comes next. Situated deep within the frontal lobes, the executive system occupies a fairly expansive area (called the cingulate gyrus) that lies below the deep fissure between hemispheres in the two frontal lobes. The more frontal part of this region is deeply involved in functions specific to reading: directing the visual system to focus on particular visual features in a given letter or word (for instance, a novice reader must pay close attention to the direction of "b" in "bear"); coordinating information from other frontal areas, particularly with regard to the semantic processing of the meaning of words (is a "bear hug" something you want or not?); and controlling the use of a particular kind of memory called working memory.

Cognitive scientists do not look at memory as one entity. What most people think of as memory—that is, our ability to recall *personal information and events that have happened to us—psychologists call episodic memory, to differentiate it from semantic memory, which refers to how we store words and facts. They also make a distinction between declarative memory (the system for retrieving the "whats" of our knowledge base, such as when the Declaration of Independence was signed) and procedural memory (the system for the "hows" of our knowledge, such as how to play the recorder, ride a bike, or hammer a nail).*

The next distinction in memory is the one most helpful in recognizing a word. Working memory is what we use when we have to hold on to information briefly, so that we can perform a task with it. This is our cognitive blackboard or scratch pad. Key to expert reading, working memory ensures that we can keep the initial visual identification of a word in mind long enough to add the rest of the information about the word (such as meaning and grammatical use).

When fluent readers identify a string of words, particularly one with considerable semantic and grammatical information, they use both working memory and associative memory. The latter helps us recall information that has been stored long-term, such as our first bicycle, our first kiss, or memorable other firsts!

Between 50 and 150 Milliseconds: Recognizing a Letter and Changing a Brain

A critical step in learning to read involves mastering the perceptual properties of written language, so that the visual system can talk effectively to the language system. The product of this learning is a new set of computational structures in the prestriate visual cortex that did not exist prior to reading.
— THOMAS CARR

Learning to read changes the visual cortex of the brain. Because the visual system is capable of object recognition and specialization, the expert reader's visual areas are now populated with cell networks responsible for visual images of letters, letter patterns, and words. These areas function at tremendous speeds in the expert reader, thanks to several very important processing principles, some of which were described by the twentieth-century psychologist Donald Hebb. Hebb proposed the notion of cell assemblies, groups of cells that learn to operate as working units. If a common letter pattern or a word like "bear" appears to an expert reader, it will trigger its own network, rather than individually activating the large number of unrelated individual cells responsible for the lines, diagonals, and circles within its letters. This operating principle is the working example of the biological maxim "Cells that fire together stay together," and is the brain's basic tool for creating ever larger circuits that connect cell assemblies into a system of networks distributed across the entire brain. The expert reading brain is a veritable collage of these networks, for every type of mental representation across the entire brain, from visual and orthographic pattern representations to phonological ones. As we saw earlier in Stephen Kosslyn's research with imagined letters, we can retrieve these representations at lightning speed even when the initial stimulus is not actually before our eyes but is only in the mind's eye.

Another contribution to automaticity involves the seemingly simple way our eyes move across text. This may appear smooth and effortless, but as Keith Rayner, an expert in eye movements, points out, that is just an illusion. Research reveals that our eyes continually make small movements called saccades, followed by very brief moments when the eyes are almost stopped, called fixations, while we gather information from our central (foveal) vision. At least 10 percent of the time, our eyes dart back ever so slightly in regressions to pick up past information. When adults read, the typical saccade covers about eight letters; for children it is less. One brilliant design feature of our eyes allows us to see "ahead" into a parafoveal region and still farther along the line of text into the peripheral region. We now know that when we read in English, we actually see about fourteen or fifteen letters to the right of our fixed focus, and we see the same number of letters to the left if we read in Hebrew.

Because we use foveal and parafoveal information, we always have a preview of what lies ahead. The preview then becomes—

milliseconds later—easier to recognize, contributing further to our automaticity. As Rayner describes, most amazing in this realm of eye movements and their rules is the closeness of the connection between eye and mind.

This link is observable. If you look at the time line, you see that many visual and orthographic representational processes happen between 50 and 150 milliseconds; and then, sometime between 150 and 200 milliseconds, the executive and attention systems of the frontal lobes activate. This is when our executive system influences the next eye movements. The executive system determines whether there is enough information about letters and word forms to move forward to a new saccade at 250 milliseconds, or whether a regression backward is needed to get more information.

Another contribution to automaticity in the sequence of our eye movements concerns our ability to recognize when a group of letters forms a permissible pattern in our language (bear versus rbea), and whether a permissible word is a real word or not (bear versus reab). At about 150 milliseconds on the time line some important occipital-temporal areas (referred to by neuroscientists as area 37) become important. As discussed briefly earlier, the researchers Stanislas Dehaene and Bruce McCandliss argue that when a child acquires reading, some neurons in this area learn to become specialized in the orthographic patterns of the particular writing system. Their hypothesis is that this ability evolved from object-recognition circuits. If so, Victor Hugo's observations about the natural origins of letters and characters—Y's and rivers, S's and snakes, C's as crescent moons—would be not only fascinating but prescient. Dehaene and his group argue that the same areas used for recognizing snakes, plows, and moons come to be used for recognizing letters. These changes in visual specialization reach a zenith in the expert reader, who is equipped with circuits in the visual cortex that did not exist prior to reading. These changes underlie one of the major ways that literacy has changed the human brain. So far so good.

Dehaene's group, however, goes on to hypothesize something more controversial—that these specialized populations of neu-

rons in the occipital-temporal region in area 37 become a "visual word form area," which allows the reader to know whether any group of letters constitutes a real word or not, somewhere around 150 milliseconds. A cognitive neuroscience group in England disagrees, and presents a still more complex scenario. By using time-sensitive brain imaging technology, MEG, that depicts when various structures are activated during the first milliseconds, they've found that even before area 37 brings information about a word's form to consciousness, frontal areas may be mapping the letter information into phonemes. It remains to be seen whether these activated frontal areas actually engage in phonological mapping or plan for it, since these areas might also be involved in executive functions. But the near simultaneity of the first processes in expert reading shown in these MEG images is remarkable. Whether either group is correct, together they underscore the rapid feedback and feed-forward mechanisms at work every time the brain reenacts the alphabetic principle, in the next 100 to 200 milliseconds.

100 to 200 Milliseconds: Connecting Letters to Sounds and Orthography to Phonology
Knowing the rules of a given language for letter-sound or grapheme-phoneme correspondence is the essence of the alphabetic principle, and becoming expert in these connections changes the way the brain functions. The person who hasn't learned these rules has a different brain by adulthood, a brain that is less precisely attuned to the sounds of his or her own language. An intriguing set of studies by Portuguese researchers highlights just how different the brain becomes on the basis of literacy. They studied people in remote rural Portugal who, for social and political reasons, had never had an opportunity to attend school. They compared this group with a similar group of people in rural areas who had managed to acquire some degree of literacy later in their lives, and found behavioral, cognitive-linguistic, and neurological differences between these groups. On linguistic tasks that elicit how able we are to perceive and understand the phonemes of our language (for example, try saying "birth" without

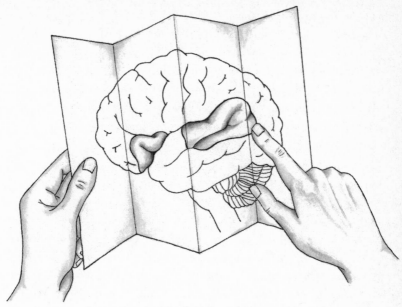

Figure 6-5: Phonological Map

the "b" sound), only the literate individuals were able to detect phonemes in speech. Becoming literate had helped them understand that words consist of sounds, which can be broken up and rearranged. On being asked to repeat nonsense words (such as "benth"), the nonliterate subjects were not readily able to do so, and would try to transform the nonsense word into a similar real word (like "birth").

Later brain scans of these two groups, done when they were in their sixties, revealed even greater differences. The brains of the people in the nonliterate group handled the language tasks with areas in the frontal lobes (as if these were problems to be memorized and solved), whereas the literate group used language areas in the temporal lobe. That is, rural people, who had been raised similarly, processed language very differently in their brains, according to whether or not they had become literate. Learning the alphabetic principle changed the way the brain performed not only in the visual cortex, but also in regions underlying auditory and phonological operations such as perception, discrimination, analysis, and the representation and manipulation of speech

sounds. *The explosion of current research on phonological pro-*
cesses shows great anatomical activity for these processes between
150 and 200 milliseconds, in multiple cortical areas including
frontal, temporal, and some parietal regions (Figure 6-5), as well
as the right cerebellum.

The specific phonological skills used in reading depend on the
reader's expertise, the word to be read, and the writing system
involved. A highly regular, highly frequent word like "carpet" will
take far less phonological processes than, say, "phonological." As
we saw in earlier phases, the novice English reader painstakingly
assembles the phoneme representations of letters and learns to
blend them into a word. This process sometimes extends over sev-
eral years. By contrast, in more regular languages like German or
Italian, readers quickly learn the far more consistent letter-sound
rules and bypass almost a year of laborious decoding. This differ-
ence among alphabetic writing systems affects how the cortex de-
ploys its phonological regions in the time line. Readers in the more
regular Finnish, German, and Italian alphabets actually get to their
temporal lobe areas more quickly, and use them more extensively,
than do English or French readers. English and French readers use
the temporal regions, but they appear to employ more of the re-
gions devoted to identifying words in the putative visual word
form area. Presumably, the greater emphasis on morphemes and
irregular words (such as "yacht") in English and French requires
more visual and orthographic representational knowledge during
this period of 100 to 200 milliseconds. The same general principle
applies to Chinese and Japanese kanji readers, who recruit this left
posterior occipital-temporal region around area 37 somewhat
more than any other adult readers, as well as right hemisphere oc-
cipital areas. For Chinese readers phonological areas are less prom-
inent during this period (100 to 200 milliseconds).

200 to 500 Milliseconds: Getting to All That We
Know about a Word
Knowledge about words is always evolving, not only for the
reader but also for the scientists who study it. Some cognitive neu-
roscientists track brain electrical activity during the phases of se-

mantic processing when the varied meanings and associates of words become activated. For example, my colleague at Tufts Phil Holcomb studies what we do when we process the meanings of words in sentences that have an incongruous ending. ("The lobster swallowed a mermaid.") Using a technique called evoked response potential (ERP), he finds bursts of electrical activity between 200 and 600 milliseconds after we see an incongruous word such as "mermaid," peaking at 400 milliseconds. Research like this gives us two bits of information for the time line: first, it indicates that the retrieval of semantic information first comes in around 200 milliseconds for typical readers; second, it indicates that we continue to add information, particularly around 400 milliseconds, if there is a semantic mismatch with our predictions.

In childhood and in expert reading, the more established our knowledge of a word, the more accurately and rapidly we read it. Think about one of the rather intimidating words you came across in an earlier chapter: "morphophonemic." Before you had read this book, this word might have slowed your reading considerably. Now it elicits knowledge that accelerates your recognition and understanding. How fast we read any word is greatly influenced by the quality and quantity of the semantic knowledge we have that is activated along with the word. Just as in the earlier phases in childhood, there is a continuum of word knowledge for adults, ranging from unknown to acquainted to well established. Where a word lies along this continuum depends on its frequency (how often a given word appears in text), a person's familiarity, and recency of exposure. Think of "sesquipedalian," which, as the essayist Anne Fadiman points out, looks as if it refers to a "long word," and does. In her Confessions of a Common Reader, Fadiman provides a list of uncommon words that would test any expert reader's mettle with regard to frequency: monophysite, mephitic, diapason, adapertile, and goetic are a few that brought me to my knees. Each of Fadiman's words, which qualify for the low end of the continuum of word familiarity, would cripple our efficiency, even though each of them also has highly familiar morphemes that tease us with hope.

Finnish researchers found that the upper temporal lobe regions involved in both phonological and semantic processing activate more quickly for words on the "established" end of this continuum. And, as noted earlier, the "richer" a semantic "neighborhood" (associated words and meanings that contribute to our knowledge about a word), the faster we recognize a word. The implications of these related semantic principles apply to people of all ages: the better you know a word and the more you know about it, the faster you read it. Furthermore, having a richly connected, established vocabulary or semantic network is physically reflected in the brain: extensive distribution in the 200-millisecond to 500-millisecond time frame reflects the variety of phonological processes and elaborated semantic networks coming to bear. The more of these networks are activated, the faster the overall efficiency in the brain for reading the word.

Syntactic and Morphological Processes

Like semantic processes, syntactic information appears to be automatically utilized sometime after 200 milliseconds from frontal areas such as Broca's, from left temporal areas, and also from the right cerebellum. Syntactic processes are used most extensively with connected text (such as sentences or text passages) and often require some feed-forward, feed-backward operations (such as those you used for "the bow on the boat"), and considerable application of working memory. Words like "bear" and "bow" contain syntactically ambiguous information and need the context of a phrase or sentence to convey more information. Syntactic information is intrinsically connected both to semantic knowledge and to morphological information, and the ability of these collective systems to work together facilitates efficiency in the period from 200 to 500 milliseconds. (For example, if you know that the morpheme "ed" is a syntactic marker for the past tense, you will identify and comprehend a word like "bowed" faster.) As Figure 6-6 shows, the more we know about the underlying life of any word, the more cumulative and convergent the contributions from different brain areas are, and the better and faster we read that word.

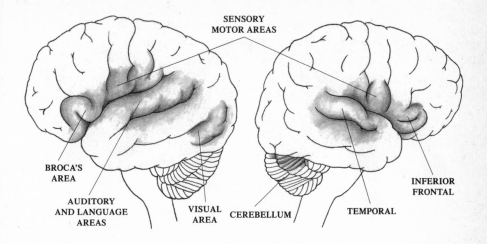

SENSORY
MOTOR AREAS

BROCA'S
AREA

INFERIOR
FRONTAL

AUDITORY
AND LANGUAGE
AREAS

VISUAL
AREA   CEREBELLUM   TEMPORAL

Figure 6-6: How the Brain Reads a Word Aloud

Once we begin to grasp what is required in order for our brains to read a single word, we can't help asking how in the world we read whole sentences and paragraphs, let alone whole books. For this we need to move outside the word's time line to consider the breathtaking achievement of reading and comprehending *Moby-Dick*, the physicist Stephen Hawking's *A Brief History of Time*, or the evolutionary biologist Sean Carroll's *Endless Forms Most Beautiful*.

HOW WHAT WE READ CHANGES US OVER TIME

Reading is experience. A biography of any literary person ought to deal at length with what he read and when, for in some sense, we are what we read.
—JOSEPH EPSTEIN

For every thinking person each verse of each poet will show a new and different face every few years, will awaken a different resonance in him. . . . The great and mysterious thing about this reading experience is this: the more discriminatingly, the more sensitively, and the more associatively we learn to read, the

more clearly we see every thought and every poem in its unique-
ness, its individuality, in its precise limitations.
—HERMANN HESSE

The degree to which expert reading changes over the course of
our adult lives depends largely on what we read and how we read
it. Such changes are best captured, perhaps, not by cognitive studies and images of the brain but by our poets. William Stafford
expressed the first element in these changes when he wrote, "A
quality of attention has been given you." He may not have been
talking about attention networks or expert readers, but this almost ineffable quality in how we attend to a text changes over
time as we learn to read—in the German novelist Hermann
Hesse's words, "more discriminatingly, more sensitively, more associatively." As we mature, we bring to the text not only all the
cognitive expertise described in the time line for words, but also
the impact of life experiences—our loves, losses, joys, sorrows,
successes, and failures. Our interpretative response to what we
read has a depth that, as often as not, takes us in new directions
from where the author's thinking left us. This explains how we
can read the Bible, *Middlemarch*, or *The Brothers Karamazov* at
ages seventeen, thirty-seven, fifty-seven, and seventy-seven and
come away with an entirely new understanding each time. I would
like to take several examples from the last two of these works to
illuminate both what we might have missed and what we know
differently based on the quality of attention and life experiences
we bring to each reading.

First, some context for the passage below. In George Eliot's
nineteenth-century novel *Middlemarch*, the beautiful, idealistic,
young heroine Dorothea Brooke cannot be dissuaded from marrying a much older man, a scholar, Mr. Casaubon. She wishes to
marry Mr. Casaubon primarily to aid him in bringing to fruition
his ambitious literary project. During their honeymoon in Rome,
Mr. Casaubon visits many libraries and Dorothea is left to her
own thoughts.

How was it that in the weeks since her marriage
Dorothea had not distinctly observed but felt with a

stifling depression, that the large vistas and wide fresh air which she had dreamed of finding in her husband's mind were replaced by ante-rooms and winding passages which seemed to lead nowhither?

George Eliot uses a series of metaphors in this passage to help us gradually infer that Dorothea has seen through Mr. Casaubon with his encyclopedic notes, and now knows that Mr. Casaubon has no great unifying work, no book, nothing beneath the endless, unconnected minutiae of thoughts preserved on his little white note cards.

This single sentence from *Middlemarch* illustrates several dimensions in expert reading. First, if the reader misses its implicit meaning, much of the nuances in the next fifty pages will also be missed. The metaphors here show how critical our "quality of attention" is to understanding the layers of meaning that lie within a text. Without this dimension, we would miss the real meaning of Dorothea's plight. Second, this particularly nineteenth-century sentence illustrates how important familiarity with varied syntactic structures can be for comprehension, and also how syntactic forms can reinforce an intended meaning. Eliot strings together four clauses and six phrases in this sentence before she leaves us "nowhither." It is almost as if she uses the recursive potential of syntax to re-create the endless anterooms that characterize poor Mr. Casaubon's mind. By its end this sentence's combination of syntactic demands and metaphoric language leads our attention to far deeper inferences about Dorothea's reality, inviting our deeper identification with her.

A second, later passage, this time from Mr. Casaubon's perspective, may be less memorable, and for good reason:

He had formerly observed with approbation her capacity for worshipping the right object; he now foresaw with sudden terror that this capacity might be replaced with presumption—that which sees vaguely a great many fine ends and has not the least notion of what it costs to research them.

I have read *Middlemarch* half a dozen times. Only when I read it last year did I see this passage about Mr. Casaubon in a different light. For three decades I identified completely and solely with the disillusionment of the idealistic Dorothea. Only now do I begin to fathom Casaubon's fears, his unmet hopes, and his own form of disillusion at not being understood by the youthful Dorothea. I never thought I would see the day when I empathized with Mr. Casaubon, but now, with no small humility, I admit that I do. So also did George Eliot, perhaps for reasons similar enough to my own. Reading changes our lives, and our lives change our reading.

To illustrate the intellectual processes that must come together for the highest forms of expert reading, I turn now to one of the most difficult passages from one of the world's most beautiful books, Dostoyevsky's *The Brothers Karamazov*. In the middle of this deeply probing Russian novel, the cynical Karamazov brother, Ivan, relates a terrible tale of good and evil called "The Grand Inquisitor" to his gentle, unworldly younger brother, Alyosha. This story within a story presents a charged dialogue set in the midst of the dreaded Inquisition. In the dialogue, a ninety-year-old monk caustically interrogates a deity referred to only as "You," "He," and "Him." See for yourself all the demands Dostoyevsky places on his reader and observe what you yourself must bring to the task of trying to understand this dialogue, where the monk rebukes a silent "Him" and tells Him why He must now die.

> It is this demand for a *universality* of worship that has been the chief torment of each and every man individually and of the whole of humankind from the beginning of time. For this universality of worship, men have put one another to the sword. They have created gods and appealed to one another, "Leave your gods and come and worship ours, otherwise death to you and to your gods!" . . . You knew, You could not but know, this fundamental secret of human nature, but You rejected the one absolute banner that was offered to You to make all men worship You uniquely . . . and You rejected it in the name of freedom and the bread of heaven. Look what

You have done since then. And again, all in the name of
freedom! Instead of taking control of human freedom,
You intensified it and burdened man's spiritual domain
with its torments for ever. You desired man to have
freedom of choice in love so that he would follow You
freely, lured and captivated by You. Instead of the old
immutable law, man should henceforth decide with a free
heart what is good and what is evil. . . . They could not
have been left in worse confusion and torment than that
in which You left them, bequeathing them so many
problems and unresolved questions.

Consider the lengths you just went to, simply to understand, first,
what the monk is really saying; second, why Ivan relates this to
Alyosha; and third, how an innocent Alyosha might respond to
this view of good and evil that contorts one's preconceptions. Before you began to read a single word, the contextual information I
provided evoked a set of executive processes for prediction, antici-
pation, and planning. These processes primed you with a particular literary genre (Russian novel) and historical setting (a dialogue
between a monk and a divine presence during the Inquisition).
Next, as you decoded the text, you placed the surface representations of words in temporary storage (working memory) to "hold"
highly sophisticated knowledge—not only about the meanings of
individual words and phrases ("universality of worship") and their
grammatical uses, but also about a number of difficult, sometimes
counterintuitive propositions in the text (worship as torment; freedom as torture; freedom of choice as lure). Meanwhile, meanings
for these concepts activated long-term memory for general background knowledge—of nineteenth-century Russia, of the Inquisition, of philosophical thinking about goodness and evil, and of
Dostoyevsky's use of the novel for didactic purposes.

Next, in all likelihood, you began to infer possible meanings
and generated a series of hypotheses about relations between Ivan
and Alyosha, the Inquisitor and Him, Dostoyevsky and his reader.
For example, you probably constructed alternative hypotheses
about what the monk was really saying and why. Throughout the

passage, you monitored your comprehension to be sure that your inferences matched your stored background knowledge. If any mismatch occurred between what was read and what was inferred, you would reread to revise your comprehension of the aberrant part or the whole.

The entire range of complexity in any text affects the comprehension of the expert reader (Figure 6-7)—from word meanings and syntactic demands to the number of conceptual propositions to be held in memory. As illuminated in this excerpt, intellectual flexibility comes to the fore to make sense of concepts that run counter to conventional assumptions (e.g., freedom as a negative value; monks who would condemn and persecute the deity). As we saw in the passages from *Middlemarch*, comprehension is affected by everything that the reader brings to the text. Ivan and Mr. Casaubon may not improve with age, but we understand them more at thirty-seven, fifty-seven, or seventy-seven than we do at seventeen.

The dynamic interaction between text and life experiences is bidirectional: we bring our life experiences to the text, and the text changes our experience of life. Few writers have better captured this interwoven relationship than Alberto Manguel in *A History of Reading*: the entire book is a history of how he and the text change the other. Sometimes we emerge after this immersion into other worlds of thought, like Manguel, with an expansion of our capacity to think, feel, and act in new and courageous ways; but wherever we are led, we are not the same.

There are physiological correlates to this experience indicating changes at the neuronal level that occur when reading reaches the expert level. Cognitive neuroscientist Marcel Just and his research team at Carnegie Mellon hypothesize that when experts make inferences while reading, there is at least a two-stage process in the brain, which includes both the generation of hypotheses and their integration into the reader's knowledge about the text. The use of these skills by expert readers corresponds to Frodo's dawning comprehension at the end of his journey about his misbegotten, hapless guide, Gollum. As Frodo sees through Gollum's twisted obsession with the Ring, he is forced first to

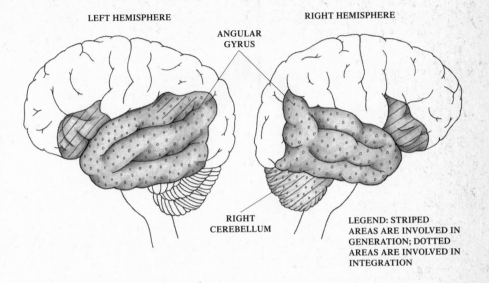

LEFT HEMISPHERE

RIGHT HEMISPHERE

ANGULAR
GYRUS

RIGHT
CEREBELLUM

LEGEND: STRIPED
AREAS ARE INVOLVED IN
GENERATION; DOTTED
AREAS ARE INVOLVED IN
INTEGRATION

Figure 6-7: Comprehension in Expert Readers

analyze and reconstruct what each action by Gollum really means, then integrate these insights into how he must proceed, and finally predict what Gollum will try next.

Like Frodo, expert readers use different comprehension processes, as well as different semantic and syntactic processes—with all their corresponding regions in the cortex—to figure out a text. For example, when readers generate inferences about what a text might mean, researchers find a bi-hemispheric frontal system activating around Broca's area. Furthermore, whenever the words used are semantically and syntactically complex, this frontal area interacts with Wernicke's area in the temporal lobe, with some parietal areas, and also with the right cerebellum. Second, and equally important, when expert readers integrate this generated inference with the rest of their background knowledge, an entire language-related system in the right hemisphere seems to be used. This second set of inferential processes requires far more work by the right hemisphere system than is needed by the earliest simple decoding tasks of the beginning reader. The right-hemisphere language system changes greatly during the development of reading and becomes as expansive and broadly distributed as the left-hemisphere language areas. Ultimately, in the expert reader there is greater left- and right-hemisphere involvement of Broca's area, as well as multiple temporal and parietal areas, including the right angular gyrus area and the right hemisphere of the cerebellum. Based on Just's research, Figure 6-7 shows that the expert reader's comprehending brain presents a beautiful change from novice reading: by using many parts of the brain, the expert reader is living testimony to our continuously expanding intellectual evolution.

. . . .

IF I COULD HAVE what Hemingway always sought—"one true sentence"—to end this natural history of the development of reading, it would be this. The end of reading development doesn't exist; the unending story of reading moves ever forward, leaving the eye, the tongue, the word, the author for a new place from which the "truth breaks forth, fresh and green," changing the brain and the reader every time.

. . . .

AS WE TURN NOW to look at the very different "natural history" of individuals with dyslexia and the ultimately hopeful genetic tale that accompanies it, we will be looking both to the preliterate past and to the future of the reading brain. We will, therefore, be traveling into uncharted territory to place the achievements of written language within a wider context, where the world of the word meets the world of the image and patterns inexpressible by speech.

PART III

WHEN *the* BRAIN CAN'T LEARN *to* READ

*For reading and writing, three years or so, from the age
of ten, are a fair allowance of a boy's time. No boy and no
parent shall be permitted to extend or curtail this period from
fondness or distaste. They must of course carry their study
of letters to the point of capacity to read and write, but
perfection of rapid and accomplished execution should not
be insisted on in cases where the natural progress within the
prescribed term of years has been slower.*

—Plato

Chapter 7

DYSLEXIA'S PUZZLE AND THE BRAIN'S DESIGN

The greatest terror a child can have is that he is not loved,
and rejection is the hell he fears. I think everyone in the world to
a large or small extent has felt rejection, and with
the crime, guilt—and there is the story of mankind. One child,
refused the love he craves, kicks the cat and hides his secret
guilt; and another steals so that money will make him
loved; and a third conquers the world—and always the
guilt and revenge and more guilt.
—JOHN STEINBECK

I would rather clean the mold around the bathtub than read.
—A CHILD WITH DYSLEXIA

JACKIE STEWART, THE SCOTTISH RACING DRIVER, won twenty-seven Grand Prix titles, was knighted by Prince Charles, and had one of the world's most successful racing careers before he retired. He is also dyslexic. Recently, he concluded a speech at an international scientific conference on dyslexia by saying, "You will never understand what it feels like to be dyslexic. No matter how long you have worked in this area, no matter if your own children are dyslexic, you will never understand what it feels like to be humiliated your entire childhood and taught every day to believe that you will never succeed at anything."

As a parent of a child with dyslexia, I know that Jackie Stewart was right. The plot of the dyslexia story is one that could be told with minor variations all around the world. A bright child, let's say a boy, arrives at school full of life and enthusiasm; he tries hard to learn to read like everyone else, but unlike everyone else he can't seem to learn how; he's told by his parents to try harder; he's told by his teachers that he's "not working to potential"; he's told by other children that he's a "retard" and a "moron"; he gets a resounding message that he's not going to amount to much; and he leaves school bearing little resemblance to the enthusiastic child he was when he entered. One can only wonder how many times this tragic story has been repeated, just because of failure at learning to read.

If a struggling young reader is lucky, however—very lucky—someone along the way will help him or her discover an "unexpected talent." Jackie Stewart said that if he hadn't discovered that he could race cars, he would surely have been "in jail, or worse," because he *had* learned how to use a gun. Only much later, after his two sons were diagnosed with dyslexia, did Stewart come to understand his own early life. He vowed that his sons would never repeat it. Late diagnosis is another frequent reality in the story of dyslexia. After their children were diagnosed with dyslexia, the financier Charles Schwab, the writer John Irving, and trial lawyer David Boies recognized their own dyslexia. Russell Cosby discovered his disability after his nephew Ennis, Bill Cosby's son, was diagnosed in college by the educator and dyslexia expert Carolyn Olivier.

Sometimes this story has a happy ending. After being asked to leave several high schools, Paul Orfalea went on to become the founder of Kinko's, David Neeleman became the CEO of JetBlue, and John Chambers became the CEO of CISCO. But a happy ending is not necessarily the norm. What frustrates me and many of my colleagues in dyslexia research is knowing that this cycle of failure can largely be avoided. We now know how to identify most children at risk of reading failure, well before they begin to experience this kind of failure, which is devastating for youngsters. When children beat their heads against a wall of failure for several years, they are often scarred for life. Jackie Stewart revealed that as an adult he could never really feel good about himself, no matter how many prizes he won or how many cars and airplanes he owned. His childhood mortification had lasted too long. Even though his is a story of resiliency, it is also an account of the terrible and lasting effects of rejection in early learning.

Examining why some brains cannot acquire written language gives us new insights into how the brain works, much as the central nervous system of a squid that can't learn to swim quickly teaches us about what is required for swimming. And vice versa: understanding the developing reading brain sheds new light on dyslexia. In the process of examining both we are invited to take a broader view of intellectual evolution—to see that a cultural invention like reading is only one expression of the brain's phenomenal potential.

. . . .

As WE EMBARK on the study of dyslexia we find very quickly that it is an intrinsically messy enterprise. There are at least three sets of reasons: the complex requirements for a reading brain; the fact that so many disciplines have been involved in its study; and the perplexing juxtaposition of singular strengths and devastating weaknesses in individuals with dyslexia. The history of dyslexia mirrors all this complexity. It also reflects many changes in our intellectual history and in our society over the last 100 years— such as Noam Chomsky's linguistic revolution and the effects of social class on the diagnosis of dyslexia. What's missing, ironically, is a single, universally accepted definition of dyslexia itself. (See the Notes for some definitions in the United States and in England, and for some of the issues involved.) Some researchers eschew the term "dyslexia" altogether and use more general descriptions such as "reading disabilities" or "learning disabilities." And despite the fact that Plato and the ancient Greeks were aware of the phenomenon, there are some who still argue dyslexia doesn't exist. I prefer the term "dyslexia," for historical reasons, but it is ultimately of no consequence what we call the brain's inability to acquire reading and spelling, as long as we understand the fascinating insights it provides and the tragic waste it can cause if unaddressed.

An Elephantine History

The tangled tale begins as it should—in our evolutionary past. Its backdrop is best captured by the British neuropsychologist Andrew Ellis who declared that whatever dyslexia turns out to be, "it is not a reading disorder." Ellis was referring to the fact that in terms of human evolution the brain was never meant to read; as we've seen, there are neither genes nor biological structures specific only to reading. Instead, in order to read, each brain must learn to make new circuits by connecting older regions originally designed and genetically programmed for other things, such as recognizing objects and retrieving their names. Dyslexia cannot be anything so simple as a flaw in the brain's "reading center," for no such thing exists. To find the causes of dyslexia, we must look to older structures of the brain and their multiple levels of processes, structures, neurons, and genes, all of which have to come together in rapid synchrony to form the reading circuit.

In other words, we must look once again, but with closer attention, at the five layers of the pyramid of reading presented earlier. Shown again in Figure 7-1, the pyramid represents the activity that supports each basic behavior at the top layer, such as reading a word or a sentence. I use it now with new intention: to help chart the various places and ways that the development of the reading circuit can go awry. The second cognitive layer of the pyramid, which consists of basic perceptual, conceptual, linguistic, attentional, and motor processes, is the plane that many psychologists study. Most twentieth-century theorists believed that difficulties within this layer were the primary explanation for dyslexia. The many processes in this layer, in turn, rest on neurological

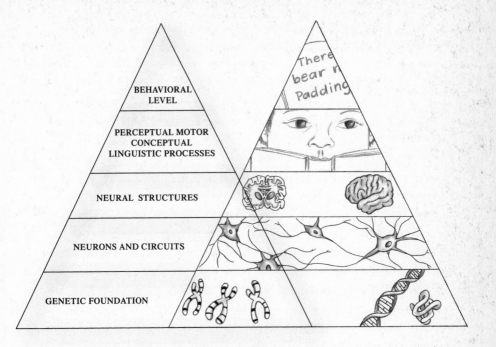

Figure 7-1: Pyramid of Reading Behaviors

structures, which—when connected—form the circuits that allow us to learn to read. A great deal of recent imaging research investigates these structures and their connections in an effort to understand dyslexia. Underlying this structural layer is a layer composed of working groups of neurons. Their ability to make and retrieve lasting representations of various forms of information allows humans to become specialists at seeing and hearing—for instance—letters and phonemes, and to do so automatically.

The pyramid's final layer represents the genes that program neurons to form working groups, structures, and ultimately circuits for older processes such as vision and language. Some of the most recent dyslexia research is concerned with these last layers. This work is complicated by the fact that the circuit for reading has no genes unique only to itself to pass on to future generations. The top four layers must learn how to form the necessary pathways anew every time reading is acquired by an individual brain. As a result, reading and other cultural inventions differ from other processes; they do not come "naturally" to children, unlike language or vision, and young novice readers are especially vulnerable.

The evolutionary perspective on the reading brain presented in this book begins with the three principles of organization that permitted the brain to read the first token. Across all written languages, reading development involves: a rearrangement of older structures to make new learning circuits; a capacity for specialization in working groups of neurons within these structures for representing information; and automaticity—the capacity of these neuronal groups and learning circuits to retrieve and connect this information at nearly automatic rates. If we apply these design principles to reading failure, a number of potential basic sources for dyslexia emerge: (1) a developmental, possibly genetic, flaw in the structures underlying language or vision (e.g., a failure of working groups to learn to specialize within those structures); (2) a problem in achieving automaticity—in retrieving representations within given specialized working groups, or in the connections among structures in the circuit, or both; (3) an impediment in the circuit connections between and among these structures; and (4) the rearrangement of a different circuit altogether from the conventional ones used for a particular writing system. Some causes of reading problems will be found across all writing systems, and some may prove relatively unique to a particular system.

Over the last 120 years of the untidy history of dyslexia research, every one of these four types of breakdowns arises in one hypothesis or another. In fact, organizing the various hypotheses on reading failure according to these principles will tidy up the history considerably. More important, by organizing the collective information provided by different theories of dyslexia along the lines of the brain's design, we can see a far clearer picture of how the study of reading failure refines our knowledge of the reading brain.

Principle 1: A Flaw in the Older Structures

The great majority of twentieth-century theories of dyslexia explains it in terms of one of the older structures in the circuit, be-

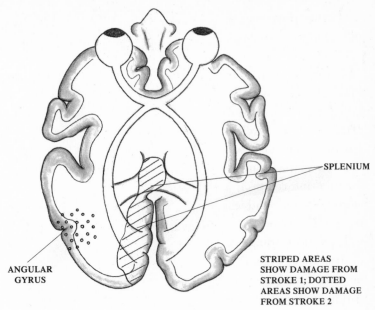

SPLENIUM

ANGULAR
GYRUS

STRIPED AREAS
SHOW DAMAGE FROM
STROKE 1; DOTTED
AREAS SHOW DAMAGE
FROM STROKE 2

Figure 7-2: Alexia Brain

ginning with the visual system. The first term for what we now call dyslexia was "word-blindness." It dates back to the work of the German researcher Adolph Kussmaul in the 1870s. Childhood dyslexia came to be called congenital word-blindness, on the basis of both Kussmaul's work and the strange case of Monsieur X—a French businessman and amateur musician who woke up one day to discover that he could barely read a word. The French neurologist Joseph-Jules Déjerine found that Monsieur X could indeed no longer read words, name colors, or read musical notes, despite having completely intact vision. After several years, Monsieur X suffered a stroke that destroyed all his ability to read and write and caused his death.

An autopsy of Monsieur X revealed two separate strokes, each of which had damaged discrete areas of the brain. Déjerine used this information as the basis for a new theory about reading and the brain. The first stroke had caused a lesion in the left visual area and at the back of the corpus callosum, the band of fibers that connects the brain's two hemispheres (see Figure 7-2). In this first stroke Monsieur X's visual areas were "disconnected," al-

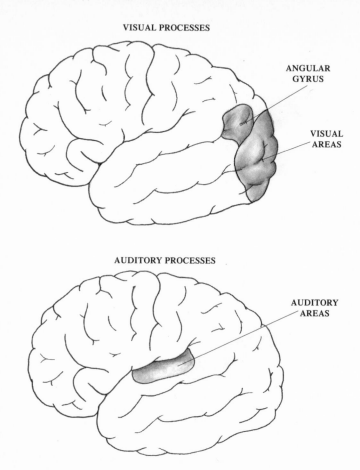

Figure 7-3: Visual Processes and Auditory Processes

lowing him to see with his right hemisphere but not to connect what he saw there either to the left-hemisphere language areas or to a damaged left visual area. This is what caused his initial inability to read. The second stroke, which caused a complete loss of all reading and writing, damaged the angular gyrus area. Déjerine's case of "classic alexia" marked the real beginning of research on acquired dyslexia and was the basis for the first hypotheses about the role of vision and the importance of connections.

The twentieth-century neurologist Norman Geschwind translated Déjerine's case as an example of a "disconnexion syndrome," which occurs when different parts of the brain necessary for a given function—such as written language—are cut off from one another, causing the function to break down. Thus Monsieur X's case actually reflects two different hypotheses: first, damage to one of the older structures, the visual system; and second, an impediment in connections in the reading circuit.

Another early, logical explanation for reading failure was a problem in the auditory system (see Figure 7-3). The reading researcher Lucy Fildes argued, in 1921, that children with problems in reading were not able to form auditory images (these are similar to our notion of phoneme representations) of the sounds represented by letters. In 1944, the neurologist and psychiatrist Paul Schilder astutely described the impaired reader as one who is not able to relate letters to their sounds, and not able to differentiate a spoken word into its sounds. Schilder's insights and Fildes's earlier work on acoustic images are the precursors of one of the most important directions in modern work on dyslexia—children's inability to process phonemes within words.

At the start of the 1970s, largely on the basis of the intellectual influence of the linguist Noam Chomsky, the emerging field of psycholinguistics (the study of the psychology of language) charted a new course for the study of reading. The aim of the early psycholinguists was nothing less than a systematic understanding of the relationships among speech, language, reading development, and reading failure. Their view that dyslexia was a language-based disorder overturned earlier, more perceptual and visual-based theories. In one of the most thought-provoking studies from this perspective, the psychologists Isabelle Liberman and Don Shankweiler studied a group of profoundly deaf children, who of course had no ability to hear speech. They found that only a small number of the children could read well, and these readers differed from the others in having a phonological representation of sounds within words. Liberman and Shankweiler interpreted these and other findings to mean that reading depended more on the linguistically demanding skills of phonological analysis and awareness (Figure 7-4) than on sensory-based auditory perception of speech sounds.

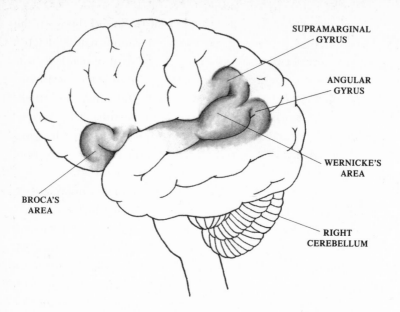

Figure 7-4: Language Hypotheses and Phonological Processing

The experimental psychologist Frank Vellutino completed the move of the field of reading disabilities away from perceptual structures as explanations of reading failure. Vellutino and his colleagues demonstrated that the most common perceptual problems in dyslexia, the well-known "visual" reversals (such as b for d, or p for q), resulted not from perceptual deficits but from the child's inability to retrieve the correct verbal labels for these sounds. In a delightfully shrewd study, Vellutino first showed impaired-reading children several typical reversal pairs (like b and d) and then asked them either to draw the letters (a nonverbal task invoking visual processes) or to say them (a verbal task). The children drew the letters very accurately, but consistently gave the wrong names, indicating a language-based source of the breakdown.

There are now hundreds of phonological studies demonstrating that many children with reading disabilities do not perceive, segment, or manipulate individual syllables and phonemes in the same way as average-reading children do. The importance of this finding was far-reaching. Children who are not aware that the word "bat" has three separable sounds will have difficulty if a well-meaning teacher begins a lesson with, "Sound the word out into its parts: \b\ - \a\ - \t\." These children cannot readily delete a phoneme from the beginning or end of a word, much less from the middle, and then pronounce it; and their awareness of rhyme patterns (to decide whether two words like "fat" and "rat" rhyme or not) develops much more slowly. More significantly, we now know that these children experience the most difficulties learning to read when they are expected to induce the rules of correspondence between letters and sounds on their own.

Indeed, the most important contribution of phonological explanations of dyslexia is their impact on early reading instruction and remediation. The researchers Joseph Torgesen and Richard Wagner and their colleagues at Florida State University have demonstrated that programs which systematically and explicitly teach young readers phoneme awareness and grapheme-phoneme correspondence are far more successful in dealing with reading disabilities than other programs. The sheer amount of evidence showing the efficacy of phoneme awareness and explicit instruction in decoding for early reading skills could fill a library wall. Phonological research thus represents the most studied structural hypothesis of reading failure.

Other somewhat less studied but nonetheless essential structural hypotheses range from the frontal lobes' executive processes—which include the organization of attention, memory, and the monitoring of comprehension—to the posterior regions of the cerebellum, which are involved in many aspects of timing, of language processes, and the connections between motor coordination and ideation. The importance of any of these structural hypotheses is twofold. Some children, as Virginia Berninger of the University of Washington demonstrates, have reading problems that stem from more primary issues in executive processes such as attention and memory; others have comorbid problems in reading and attention. As elaborated below, still others have timing-related issues. Several British researchers hypothesize these may, in at least some children, involve cerebellar dysfunction.

The overall point of this section, however, is the collective picture that results when one examines all the structural types of

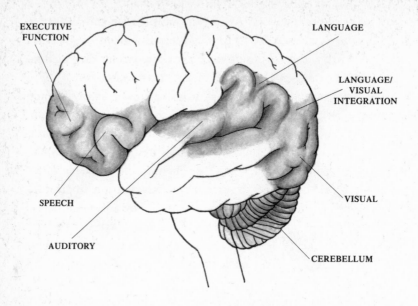

EXECUTIVE
FUNCTION

LANGUAGE

LANGUAGE/
VISUAL
INTEGRATION

SPEECH

VISUAL

AUDITORY

CEREBELLUM

Figure 7-5: Cumulative Dyslexia Hypothesis

hypotheses. Throughout the early and middle twentieth century, well-meaning researchers tended to claim one area of dysfunction and to assert that this was the primary explanation for most reading failure. Although it may well be the most overused metaphor in the field of dyslexia, the story of the blind men and the elephant remains an apt depiction of much of this research.

Not surprisingly, many theorists gave their particular explanation of reading failure a new name. Consider what occurs if we place all the historical hypotheses for failure at a process-structural level, like pieces on a map of the human brain (see Figure 7-5). Voilà: the sum of these hypotheses looks like a decent approximation of the major parts of the universal reading system. It is another way of saying that many of the collective hypothesized sources of dyslexia mirror the major component structures of the reading brain.

PRINCIPLE 2: A FAILURE TO ACHIEVE AUTOMATICITY
A second type of hypothesis highlights the failure to achieve automaticity, or sufficiently rapid rates of processing, within or

among those structures. The underlying premise is that as a result of this failure—whether at the level of neurons or structural processes—the various parts in the reading circuit do not function fluently enough to allocate time for comprehension.

As with the first set of hypotheses, there are many fluency-related explanations that address different levels of the pyramid and different structures. Unsurprisingly, several of these begin, as before, with vision. For example, Bruno Breitmeyer and the Australian researcher William Lovegrove found considerable differences in the speed of processing visual information in dyslexia. Think of an image of a star followed quickly by another image of a star. In the brain of many an individual with dyslexia, two rapidly presented visual "flickers" appear fused into one stimulus, because the person cannot process this visual information quickly enough.

Analogous research on how quickly children with dyslexia process auditory information indicates somewhat similar differences with average readers. In both processes, impaired readers are like their peers at the most basic level of detection: they readily perceive when a visual stimulus or a sound occurs. But with a little added complexity, differences appear. Some children with reading disabilities and many children with language impairments require longer intervals than their peers in dealing with two brief, separated tones, much as with visual images. Increasingly sophisticated research demonstrates that these difficulties are compounded by factors affecting the finer phonemic and syllabic distinctions within words. Usha Goswami of Cambridge Universisty, for example, found that the children with dyslexia she studied in England, France, and Finland were less sensitive to the rhythm in natural speech, which is partly determined by how the sounds in words change through stress and "beat patterns." All this can lead to poor phoneme representations, and later reading failure.

The evidence on differences in the speed of motor processes in dyslexia remains some of the most curious and may turn out to be related to Goswami's findings about speech. After observing children trying to tap out rhythmic patterns from a metronome, a noted psychiatrist, Peter Wolff of Boston, concluded that the automaticity in motoric areas becomes problematic in dyslexia when readers must put together the individual parts of a behavior into "temporally ordered larger ensembles." In other words, whether in motor functions, eye, or ear, a breakdown occurs for a number of children with dyslexia when they need to connect the components of a task accurately, serially, and rapidly, not at the most basic level of sensory processing.

The Israeli psychologist Zvia Breznitz gives this story an unusual twist. Breznitz studied children with dyslexia, using a wide spectrum of tasks over two decades, and found an extensive range of problems with processing speed. Along the way she made an unusual discovery. Like others, she found that poor readers were characterized by slower processing in each modality, but in addition, the impaired readers appeared to have a "gap in time"—what Breznitz calls an "asynchrony"—between their visual and auditory processes. It is as if the two areas most needed to make letter-sound correspondence in reading are not sufficiently synchronized for their individual information to become integrated, with implications for reading all down the line. Also observed years ago by Charles Perfetti, Breznitz's concept of an asynchrony in time remains one of the most fascinating pieces in the puzzle of dyslexia.

Indeed, one of the single best predictors of dyslexia in every language tested is a time-related task called "naming speed" that incorporates almost all the cognitive-level processes on the second tier of our pyramid. The story of naming speed goes back to the case of Monsieur X, whose rare combination of damage rendered him unable to read and also unable to name colors. From this Geschwind reasoned that the systems for naming colors and reading must use some of the same neurological structures and share many cognitive, linguistic, and perceptual processes. And from this he reasoned that a child's ability to name colors, which develops well before kindergarten, would be a good predictor of reading acquisition and failure.

The pediatric neurologist Martha Bridge Denckla of Johns Hopkins University tested this and found that readers with dyslexia can name colors perfectly well, but they cannot name them rapidly. The time it takes for the brain to connect visual and lin-guistic processes to name colors (or letters and numbers) was the predictor of who would be unable to learn to read. Denckla's dis-covery and her work with the neuropsychologist Rita Rudel of MIT became the basis of "rapid automatized naming" (RAN) tasks in which the child names rows of repeated letters, numbers, colors, or objects as fast as possible. Extensive research in my laboratory and around the world shows that RAN tasks are "one of the best predictors of reading performance" across all tested languages. This work, in turn, became the basis of a new naming-speed task, "rapid alternating stimulus" (RAS), which I designed to add more attentional and semantic processes to the RAN nam-ing requirements. If you consider that the whole development of reading is directed toward the ability to decode so rapidly that the brain has time to think about incoming information, you will un-derstand the deep significance of those naming speed findings. In many cases of dyslexia, the brain never reaches the highest stages of reading development, because it takes too long to connect the earliest parts of the process. Many children with dyslexia literally do not have time to think in the medium of print.

Deficits in naming speed, however, were never intended to ex-plain dyslexia; rather, they represent an index of some underlying problem that is impeding the speed of reading processes. Just as Geschwind suspected, we have found that the processes and struc-tures underlying naming are a subset of the major processes and structures underlying reading. Failure in any of the major pro-cesses and structures involved in naming speed—including their connections, their automaticity, or the use of a different circuit—could cause either naming or reading deficits.

An evolutionary story lurks under the surface of naming speed and contributes to the evolving story of the first reading brain. In Figure 7-6, brain images of naming speed by the neuroscientist Russ Poldrack of UCLA and our research group show something wonderfully clarifying. Just as was hypothesized earlier by other researchers, the brain in these images uses older object recogni-tion pathways in the occipital-temporal zone (area 37) to name both letters and objects. The fMRI images support these research-

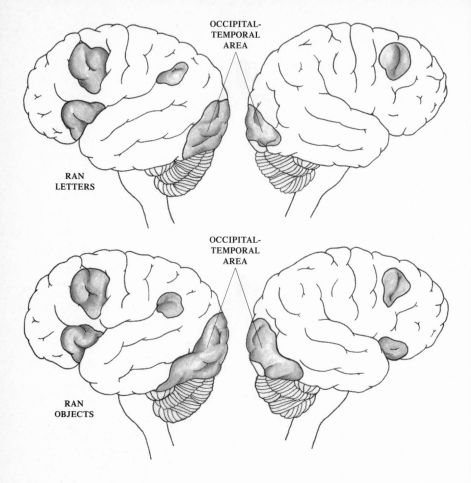

Figure 7-6: RAN fMRI

ers' hypotheses that humans are "neuronal recyclers." But a more significant story in these images involves three differences between letters and objects.

First, the important left occipital-temporal area is activated much more during object naming than during letter naming. Objects don't usually call on our capacity for superspecialization (except in interesting cases such as birds for bird-watchers), because there are so many possible objects. Thus, object recognition doesn't become totally automated, and also needs more cortical space. The object-naming circuit is a picture of us *all* before literacy.

Second, the more streamlined use of the occipital-temporal area by letters highlights the capacity of the literate brain for visual specialization and for making its specialized information automatic. This is why naming RAN letters is always faster than RAN objects for every reader.

Third, and very importantly, culturally invented letters elicit more activation than objects in each of the other "older structures" (especially temporal-parietal language areas) used for reading in the universal reading brain. This is why measures of naming speed like RAN and RAS predict reading across all known languages. It is also why, side by side, the brain images for the object- and letter-naming tasks are like comparative evolutionary photos of a pre-reading and post-reading brain.

Finally, there may be important developmental implications in the naming speed research story for the early detection of dyslexia in pre-reading children. We know that the great majority of children with dyslexia are significantly slower in retrieving names of both letters and objects early in kindergarten, and that letters then become far more predictive than objects. If object naming and letter naming represent a pre- and post-reading brain, we could look at the developing brain in children as young as age three to see if there are already weaknesses in their retrieval of object names. If we could find out early on whether a particular brain is developing an altogether different speed or even different circuit for handling object and colors—for example, if imaging research showed a very obvious difference such as right-hemisphere circuitry—we could have a far earlier predictor of future reading failure and an opportunity for earlier intervention. It is my hope that future researchers will be able to image object naming *before* children ever learn to read, so that we can study whether the use of a particular set of structures in a circuit might be a cause or a consequence of not being able to adapt to the new task of literacy.

Such complex notions move us from questions of rate and automaticity to the underlying causes of these time-related deficits. One possibility has to do with the circuit connections.

PRINCIPLE 3: AN IMPEDIMENT IN THE CIRCUIT
CONNECTIONS AMONG THE STRUCTURES

This group of hypotheses stresses the importance of understanding the connectivity among structures, rather than locating the
problem within a structure. In his translation of Déjerine's first
case of classic alexia, Norman Geschwind resurrected the concept
of the "disconnexion syndrome" from the nineteenth-century
neurologist Carl Wernicke to describe how important it is that all
component systems work together for every cognitive function.
Thus the fact that visual information in the right hemisphere
couldn't pass over the corpus callosum to the visual-verbal processes of the left hemisphere was just as important in Monsieur
X's breakdown as the structural damage in the left hemisphere.
Connections within the reading circuit are as important as the
structures themselves.

Many theorists in the mid-twentieth century emphasized this
third kind of hypothesis by considering the connections among
structures and processes in the reading circuit. The two most
common ideas located the source of failure in connections between either the visual-verbal processes or the visual-auditory
systems. Modern neuroscience goes below the surface of those
explanations to examine the functional connectivity or strength
of interactions between various structures important to reading.
Neuroscientists interested in functional connectivity investigate
the efficiency and strength of interactions among the major components in the reading circuit.

At least three forms of disconnections are consistently studied
in this type of research, and once again their cumulative information reveals a bigger story. An example of the first form of circuit
dysfunction was found by Italian neuroscientists: in Italian readers with dyslexia there appeared to be a disconnection between
frontal and posterior language regions, based on underactivity in
an expansive connecting area called the "insula." This important
region mediates between relatively distant brain regions and is
critical for automatic processing.

Researchers from Yale University and Haskins Laboratory
found a different but potentially related type of disconnection.

While studying the very important occipital-temporal region, which appears to be activated at the start of reading in any language, they found that this area 37 is not connected in the same way for readers with dyslexia. In nonimpaired readers the strongest, most automatic connections are forged between this posterior region and frontal areas in the left hemisphere. In dyslexia, however, the strongest connections appear between the left occipital-temporal area and the right-hemisphere frontal areas. In addition, some neuroscientists find that in dyslexia the left angular gyrus region that good novice readers draw on appears functionally disconnected from the other left-hemisphere language regions during reading and the processing of phonological information.

A last form of disconnection noted in imaging studies helps encompass all these findings. A research group in Houston used imaging called magnetoencephalography (MEG) that provides an approximate look at what region is activated during reading, and when. They found that children with dyslexia move from visual regions in the left and right occipital lobes to the right angular gyrus region and then to frontal areas. In other words, the children with dyslexia used an altogether different reading circuitry. These unexpected findings help to explain many mysteries, including why some of my colleagues at MIT find underactivation of the left angular gyrus region in dyslexia, and much less activation of the otherwise ubiquitous left occipital-temporal area. These findings move us from discussions of the apparent disconnections within circuits to the most provocative of the four hypotheses, the possibility of a differently rearranged brain.

PRINCIPLE 4: A DIFFERENT CIRCUIT FOR READING

Historically, the most unusual and comprehensive account of dyslexia emerged from the work of the brilliant neurologist Samuel T. Orton and his colleague Anna Gillingham. On the basis of his clinical studies in the 1920s and 1930s, Orton renamed reading disability "strephosymbolia," or "twisted symbols." Orton argued that in the brain's normal distribution of labor, the usually dominant left hemisphere selects the correct orientation of a letter (b or d) or of a letter sequence ("not" rather than "ton"). In

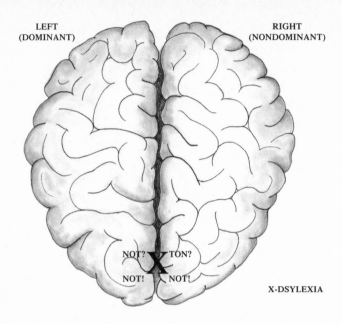

Figure 7-7: Orton's Hypothesized Strephosymbolia

dyslexia, however, this pattern of hemispheric dominance either did not occur or was dramatically delayed. As a consequence of a failure in communication between the right and left hemispheres, Orton wrote, certain children cannot select the correct letter orientation. This leads to visual spatial confusion, letter reversals, and difficulties with reading, spelling, and handwriting—that is, to dyslexia.

Researchers in the 1960s and 1970s were fascinated by the related idea that in dyslexia the left hemisphere appears weaker than the right at processing a variety of tasks having to do with reading. For example, tasks in which a child listens to stimuli presented in various ways to different ears (known as dichotic tasks) regularly showed that impaired readers were not using their left hemisphere in auditory processes in the same ways as average readers. In 1970, neuropsychologists at the Boston VA Hospital tested average readers and a group with dyslexia on a series of visual, auditory, and motor tasks. Not only was the speed of the impaired readers significantly worse for each of these tasks, but on the dichotic listening task, readers with dyslexia showed right-hemisphere superiority.

Similarly, in the 1970s researchers found an unexpected symmetry in the visual areas of persons with dyslexia during word recognition tests, with the left hemisphere surprisingly weaker at handling linguistic information. One lateralization study after another during this period showed unusual reliance on the right hemisphere in dyslexia for a range of tasks. For many years these findings were considered the product of an oversimplified view of right-brain and left-brain processing but, as we will see shortly, imaging researchers are beginning to reconsider both Orton's ideas and these older theories about hemispheric processing.

In ongoing studies of the neural development of typical reading, the research group at Georgetown University found that over time there is "progressive disengagement" of the right hemisphere's larger visual recognition system in reading words, and an increasing engagement of left hemisphere's frontal, temporal, and occipital-temporal regions. This supports Orton's belief that during development the left hemisphere takes over the processing of words.

Once again, however, this progressive development of a reading circuit is not seen the same way in dyslexia. Researchers at Yale, led by Sally and Bennett Shaywitz, first observed an unexpected circuit at work in children with dyslexia on a continuum of reading-related tasks from simple visual to more complex rhyming tasks. These children used more frontal regions and also showed much less activity in left posterior regions, particularly in the developmentally important left-hemisphere angular gyrus. Most important, this group found potentially compensatory "auxiliary" right-hemisphere regions performing functions usually handled by the more efficient left-hemisphere areas. In more recent work the Yale team studied non-impaired adults and two groups of adults with reading impairments, one that was compensated for accuracy but still dysfluent, and one with uncompensated, persistent, potentially environmental-influenced deficits. Surprising to all, the basic circuitry for the non-impaired and the uncompensated, more environment-based readers followed similar lines. The compensated readers, who were closer to the classic dyslexia profile, used more right-hemisphere regions, including the occipital-temporal regions, and underactivated the left posterior regions used by the other two groups. Furthermore, the readers with persistent deficits used the left occipital temporal region even more than the non-impaired, suggesting greater use of memory strategies than analytic ones in this group.

To whet your appetite for research to come in the near future, the artist Catherine Stoodley has created a composite sketch of some major brain-image findings about how people with dyslexia process visual, orthographic, phonological, and semantic information. The pattern in Figure 7-8 reveals something that is now entirely predictable from work on automaticity and fluency in dyslexia: delays at every step of processing from visual-orthographic recognition to semantic processing. At no point from at least 150 milliseconds on are dyslexic readers on time. In addition, something that not too long ago would have been quite astonishing is also displayed here. Readers with dyslexia appear to use brain circuitry different from that of a typical reader. The dyslexic brain consistently employs more right-hemisphere structures than left-hemisphere structures, beginning with visual association areas and the occipital-temporal zone, extending through the right angular gyrus, supramarginal gyrus, and temporal regions. There is bilateral use of pivotal frontal regions, but this frontal activation is delayed.

This time line is a product of cumulative research at many laboratories in various parts of the world, including the United States, Israel, and Finland. It is hardly finished. At best, it is thought-provoking; at worst, it is misleading. In imaging and educational research, bear in mind that Socrates' warning about text applies equally to brain images. "Their seeming impermeability gives the illusion of truth," when, in fact, they are simply our best interpretation of statistical averages on the number of subjects we have to date. Only time, and more evidence, will tell what is the truth about a different hemisphere's capacity. But should the emerging concept of a right-hemisphere-dominated reading circuit with dyslexia prove correct for some readers, then not only is

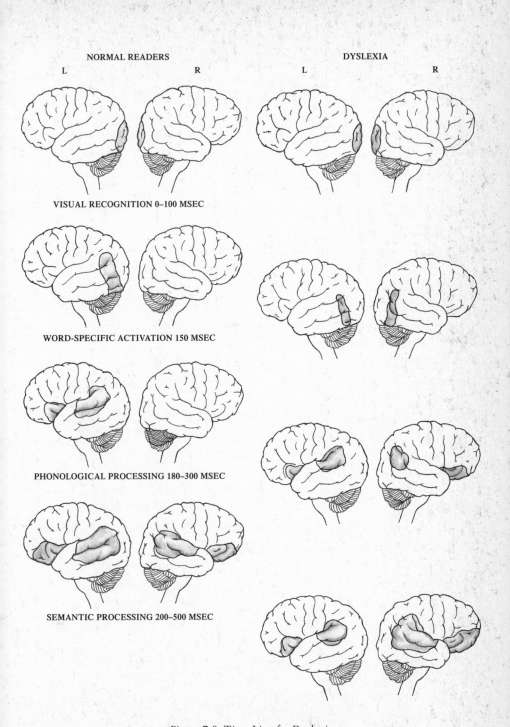

NORMAL READERS   DYSLEXIA

L   R   L   R

VISUAL RECOGNITION 0–100 MSEC

WORD-SPECIFIC ACTIVATION 150 MSEC

PHONOLOGICAL PROCESSING 180–300 MSEC

SEMANTIC PROCESSING 200–500 MSEC

Figure 7-8: Time Line for Dyslexia

the dyslexic brain in these children seeing, hearing, retrieving, and integrating orthographic, phonological, semantic, syntactic, and inferential processes more slowly; it is also doing all this with a largely different circuit of structures, in a hemisphere never designed for temporal precision.

As eminent researchers Ovid Tzeng and William Wang observed years ago, the left hemisphere evolved to handle the exquisite precision and timing necessary for human speech and written language; by contrast, the right hemisphere became better suited for operations on a larger scale, such as creativity, pattern deduction, and contextual skills. The evocative picture of a right-hemisphere-dominated circuit could help explain a century of different hypotheses, each of which accurately described one manifestation of a broader syndrome. In the context of the "reading pyramid," and organized by the basic principles of brain design presented here, the history of hypotheses about dyslexia suggests an overarching insight—that no one hypothesis will ever explain all possible forms of reading disabilities, especially across different languages.

This brings us to pressing questions about dyslexia today, and the issue of heterogeneity among impaired readers, not only in different languages but also in the same writing system. An understanding of the principles of brain design in reading moves us away from any one-dimensional account of reading disabilities, however worthy, into a multidimensional view of reading disabilities. There are various possible causes of reading failure—with all the difficult implications this fact has for intervention. It shifts the focus of research from finding the "primary cause" of dyslexia to finding the most prevalent subtypes of readers with dyslexia.

THE UNWANTED PRINCIPLE: MULTIPLE STRUCTURES, MULTIPLE DEFICITS, AND MULTIPLE SUBTYPES

Accepting the idea of subtypes is much easier than fitting real children, with their mix of characteristics that change over the course of development, into any empirically based classification system. My Canadian colleague Pat Bowers and I took an intentionally simple approach to considering multiple deficits. We studied whether children with reading problems fell into subtypes based on deficits in the two best predictors of dyslexia—subtype 1, phoneme awareness problems (a structural hypothesis); subtype 2, slow naming speed (a proxy for processing speed and fluency); and subtype 3, both deficits. About one-fourth of English-speaking poor readers had only phonological deficits. Very important, just under 20 percent of the poor readers had only fluency deficits. This "fluency-only" subtype of dyslexia, although a relatively small group in English, is a much larger group in languages such as German and Spanish that have a more regular writing system. In English, this fluency subtype is exemplified by Luke in Chapter 6, who could not read fast enough to sing his arias, but was not considered reading-impaired by his teachers. Such students are routinely missed in most schools because they begin with no real decoding problems and only later show fluency deficits and poor comprehension.

The most common, and most difficult, subtype of reader that we found in English was subtype 3: the child with double deficits in naming speed and phonology, accompanied by the most severe impairments in every aspect of reading. With both structural and processing speed deficits, these children are the ones historically described as having classic dyslexia.

Intriguingly, about 10 percent of poor readers could not be classified in this way. As described by the psychologist Bruce Pennington, this suggests the need for more encompassing multiple-subtype classifications that can someday link structural and genetic data. In one such sophisticated analysis, Robin Morris's group at Georgia State demonstrated that the most impaired group of children with dyslexia exhibit not only our combined deficits but also a deficit in short-term memory.

Until all our subtypes become more comprehensive, we have learned some useful things using the internationally transitional double-deficit framework in several dialects and language systems. For example, the proportions of children in each subtype appear similar in most English studies, but children who speak a dialect different from the standard American English dialect vary considerably. Our research group found very unusual differences among African-American impaired readers, who were matched in every way with European-American children for intelligence, instruction, and socioeconomic status. There were far more African-American children in the double-deficit and phonological subtypes, and they were disproportionately represented in the reading-disabilities population.

One promising hypothesis concerns the use by many African-American children of African-American Vernacular English (AAVE), one of several dialects in the English language. The sociolinguist Chip Gidney at Tufts and our research group are working to understand the subtle differences between standard American English and AAVE. We want to know whether the differences pose an impediment when children, long accustomed to their first dialect, try to learn grapheme-phoneme correspondence rules in a second dialect. We hope to understand whether the very subtlety of dialect differences poses a greater problem in a child's phoneme awareness than if a child speaks a language with completely different phonemes, such as Spanish or French.

What we know with more certainty is that children who use AAVE appear to have more phonological problems. In this they differ dramatically from children who speak different languages such as Spanish or Chinese. This brings us back to more universal issues about the design of the reading brain, and how dyslexia manifests itself in different languages.

Legasthenie, Dyslexi, Dyslexie: The Many Faces of Dyslexia around the World

With his German-accented but perfect English, the Austrian psychologist Heinz Wimmer sounds a lot like Henry Kissinger when he describes very real differences in how dyslexia manifests itself in German, in Dutch, and in other orthographies. Depending on what is emphasized in any given language (fluency in German; visual spatial memory in Chinese; phonological skills in English), there will be somewhat different faces of dyslexia, as well as different predictors of reading failure. As we saw in the evolution of the reading brain, different writing systems make somewhat different uses of the major structures involved in the reading circuit. It is no coincidence, therefore, that dyslexia in China has a slightly different nature. Researchers in Hong Kong found several subtypes in Chinese-speaking dyslexic children similar to our double-deficit subtype, but with a fascinating additional subtype whose major deficit is, unsurprisingly, in orthographic processes.

Among Spanish-speakers, researchers in Madrid found subtypes similar to our double-deficit classification, with one striking difference: comprehension among the most affected subtype appeared far less impaired in Spanish readers with dyslexia than in English readers with dyslexia. Similar data emerged for Hebrew. In a comparison of Hebrew-speakers and English-speakers carefully matched in every way, researchers in Haifa found that Hebrew readers were less impaired in comprehension. It appears that the shorter time needed for decoding in these languages allows more time for comprehension than in English.

The moral of these cross-language studies is that the particular emphases of a writing system influence how it breaks down. When phonological skills play a more significant role in reading acquisition, as they do in less regular languages like English and French, phoneme awareness and decoding accuracy are often very deficient—and are good predictors of dyslexia. When these skills play a less dominant role in reading (in the transparent orthographies like German, and the more logographic writing systems), processing speed becomes the stronger diagnostic predictor of reading performance, and reading fluency and comprehension issues dominate the profile of dyslexia. In these more transparent languages—Spanish, German, Finnish, Dutch, Greek, and Italian—the child with dyslexia exhibits fewer problems decoding words and more problems reading connected text fluently with good comprehension.

. . . .

THE CUMULATIVE RESEARCH of a century, organized according to principles of the brain's design for reading development and across different dialects and languages, gives us a very important window on the reading brain. It pushes us beyond what we learned from the evolution of writing systems and the development of reading acquisition in the child. It shows that everything matters in reading: the tiniest feature detectors in visual and auditory processes; the different amounts of time needed to connect the various processes in different writing systems; the question of which hemisphere does what.

Prepared with all this knowledge, researchers in the twenty-first century are beginning to ask whether the range of findings in the elephantine history of dyslexia ultimately rests on a fairly limited set of genes that govern the development of the older structures and their ability to work together proficiently. These hypotheses, to be elaborated in Chapter 8, may ultimately point to a synthesis of all four hypotheses, in which a few unusual genes cause several aberrant patterns of neuronal development in structures necessary for reading, resulting in the creation of whole new, less efficient circuits that were never meant to read.

A Century's Mystery

One hundred years ago almost no one knew that dyslexia existed. Around this time, my great-great-grandfather pushed a wheelbarrow to Indiana and built a small economic empire. As described in a history of nineteenth-century southern Indiana, he shipped millions of pounds of tobacco a year to England, despite a very interesting characteristic: "It is said that Mr. Beckmann could neither read nor write. Instead of ciphers he would make as many strokes as he had units in his accounts. Sometimes he would use ciphers, but in doing that he would get them mixed up, thus 10 would become 01." I will never know how my ancestor felt about his inability to read and his tendency to reverse numbers, but I would bet that there were moments when he felt much like Jackie Stewart—frustrated, and perhaps diminished, despite all his material success.

Fortunately, we have come to a point in time where severe reading disabilities are a familiar part of what every teacher experiences in the classroom. Our knowledge about how to predict reading breakdown has begun to inform the practice of teaching. Jackie Stewart, Paul Orfalea, Russell Cosby, and many others provide eloquent testimony about the gap between knowledge and application that affected their lives. Still, too few teachers know much about the history of dyslexia, and fewer still are aware of current trends. If I were given five minutes to speak to all teachers and parents everywhere, I would summarize the implications of the complicated twentieth-century history of dyslexia in this way:

- Learning to read, like Red Sox baseball, is a wonderful thing that can go wrong for any number of reasons. If a child cannot seem to learn to read, for no obvious reason (like abnormal vision or lack of appropriate reading instruction), it is critical to have the child evaluated by reading specialists and clinicians.
- There is no one form of dyslexia; instead, there is a continuum of developmental reading disabilities that reflects the many components of reading, as well as the specific writing system in a given language. Thus reading-impaired children can and do exhibit a variety of deficits. Some of these are subtle and involve only fluency and comprehension later in school, but at least among English-speakers, most children begin with decoding problems and an inability to learn the rules of grapheme-phoneme correspondence. This deficit often shows itself in spelling and writing as well.
- Two of the best-known deficits involve processes that underlie phonology and reading fluency. Measures of phoneme awareness and naming-speed processes, therefore, are our two best predictors of reading failure, along with vocabulary, across many languages. Children with phonological deficits typically have difficulties with the rules of letter-sound correspondence and with learning to decode. Measures of phoneme awareness will identify these children in kindergarten and first grade. By contrast, children who have only

fluency issues often exhibit early naming-speed deficits. These children are frequently overlooked because their decoding is adequate, albeit slow. As older students or adults, they experience difficulty when the amount of required reading overwhelms their slower reading rate. They are much like children with dyslexia who speak more regular languages, such as German and Spanish, and often manifest only fluency and comprehension problems. Measures of rapid naming like RAN and RAS will predict most of these children in kindergarten and first grade. Children with deficits in both phoneme awareness and naming speed require intensive intervention from the outset. A small group of children are impaired in reading but do not have problems with naming speed or phonology, and we need to know more about them.

- Some young children with severe reading disabilities come from such linguistically impoverished backgrounds that vocabulary plays a critical role. Some children who are learning English or who speak an English dialect (such as AAVE or Hawaiian pidgin) different from the one in the classroom may manifest reading disabilities that are based largely on the learning of a second language or dialect. They do not process English phonemes in the same way. It is essential to discover whether they have a reading disability in addition to learning standard American English, or whether their poor reading is based only on second-language or dialect issues.
- Intervention for children with dyslexia should address the development of each of reading's contributing components— from orthography and phonology to vocabulary and morphology—their connections, their fluency, and their integration in comprehension.
- Children with any form of dyslexia are not "dumb" or "stubborn"; nor are they "not working to potential"—the three most frequent descriptions they endure. However, they will be mistakenly described in these ways many times by many people, including themselves. It is vital for parents and teachers to work to ensure that all children with any form of

Figure 7-9: Ben Noam's Drawing of the Tower of Pisa at Age 17

reading problem receive immediate, intensive intervention, and that no child or adult equates reading problems with low intelligence. A comprehensive support system should be in place from the first indication of difficulty until the child becomes an independent, fluent reader, or the frustrations of reading failure can lead to a cycle of learning failure, dropping out, and delinquency. Most important, the considerable potential of these children will be lost to themselves and to society.

A case in point is my first son, Ben. A century after his maternal great-great-great-grandfather struggled to read, Ben struggled to read, although—like many other children with dyslexia—he had considerable intelligence and talents, and involved parents. One of the most poignant moments in the writing of this book occurred as I wrote about the perplexing lateralization hypotheses of Samuel T. Orton. As he sometimes did when he was in high school, my son sat at the dining room table beside me, drawing, as I wrote about why Orton was probably wrong at the time. I looked up to see Ben drawing with exquisite precision and detail the entire leaning Tower of Pisa—upside down (Figure 7-9)! When I asked him why, he said it was easier for him to do it that way. None of us who conduct research can adequately explain such phenomena on the basis of present knowledge. There is much we know and much that remains to be explained in the history and mystery of dyslexia. Still unresolved are the highly provocative findings about the possibility of a right-hemisphere-dominated reading circuit that could help explain Ben's different spatial capacities.

Last year, when Ben turned eighteen and was about to go off to the Rhode Island School of Design, I decided to discuss this whole line of speculation with him. We drew flowchart diagrams first of how the brain uses each hemisphere in typical readers and for what broad purposes, then of how pathways get strengthened and more automatic with their use over time, and finally of how the circuit pathways might differ dramatically in dyslexia. My husband, Gil, and I are long accustomed to being surprised by Ben; nevertheless, his first questions jolted me. "So does this mean I'm more creative because I use this right hemisphere more than other people and my right pathways got strengthened that way? Or does it mean that dyslexics are just born with more creative brains from the start?" I don't know the answer to Ben's questions. I do know that they are intimately connected to questions repeated throughout much of the new research about whether right-hemisphere reading circuits are the cause of not being able to name letters and read words easily, or the consequence.

We in the twenty-first century have an unfolding mystery at our fingertips. Because we are piecing together well-known and neglected clues from dyslexia's past history with new information from recent imaging studies, a far more comprehensive understanding is beginning to emerge of what is going on when the brain can't learn to read. I don't yet know the end of the story unraveling in the new work on dyslexia, and as a researcher I'm not altogether comfortable writing about my hunches. But if I am correct, dyslexia will turn out to be a stunning example of the strategies used by the brain to compensate: when it can't perform a function one way, it rearranges itself to find another, literally. The question of why this is so leads us to the last two layers of the pyramid and to intriguing issues of our genetic makeup.

Chapter 8

GENES, GIFTS, AND DYSLEXIA

*"The letters float off the page when you read, right?
That's because your mind is hard-wired for ancient Greek,"
explains a fellow camper, gray-eyed Annabeth. "And the A.
D.H.D.—you're impulsive, can't sit still in the classroom. That's
your battlefield reflexes. In a real fight, they'd keep
you alive. As for the attention problems, that's because you see
too much, Percy, not too little. Your senses are better than
a regular mortal's. . . . Face it. You're a half-blood."*
— RICK RIORDAN

*If only we knew
as the carver knew . . .
how the flaws in the wood
led his searching chisel
to the very core.*
— DAVID WHYTE

THOMAS EDISON, LEONARDO DA VINCI, AND ALBERT Einstein are three of the most famous people said to have had dyslexia. Thomas Edison's childhood difficulties with reading, along with his ill health, often made it impossible for him to attend formal school. Yet he went on to receive the largest number of patents granted to any one person by the United States Patent Office and to create astonishing inventions, one of which literally lit up the world.

Leonardo da Vinci was one of the most creative people in history: an inventor; a painter; a sculptor; a musician; an engineer; and a scientist. Although he was extraordinary at everything he tried, he is often considered to have been dyslexic; this conclusion is based largely on his bizarre, voluminous notes. Written from right to left, in reversed "looking-glass script," they were full of misspellings, syntactic mistakes, and strange errors in language. Several of his biographers mention his discomfort with language, as well as his frequent references to his lack of reading ability. In a poignant description of the ideal life of a painter, Leonardo wrote that it would always include a person nearby who could read to him. The neuropsychologist P. G. Aaron makes a compelling case that Leonardo's issues with reading and writing were a product of a powerful "right-hemisphere compensatory mechanism."

Albert Einstein did not speak much until three years of age, and he was mediocre at any subject that required the retrieval of words, such as a foreign language. He once said, "My principal weakness was a bad memory, especially a bad memory for words and texts." He went so far as to say that words did "not seem to play any role" in his theoretical thinking, which came to him through "more or less clear images." Whether Einstein might have met the criteria for some form of dyslexia, as he himself and Norman Geschwind believed, is unknown. But what a twist it would be if the theorist who transformed our understanding of time and space turned out to have had a timing deficit. One clue to this mystery may be his brain. Canadian neuroscientists conducted a fascinating but still disputed autopsy of Einstein's brain and discovered unexpected symmetries between the hemispheres in his enlarged parietal lobes, rather than the more typical asymmetric pattern.

Most people with dyslexia do not have spectacular talents like those of Edison or Leonardo, but there seem to be any number of persons with dyslexia who are unusually talented. I once kept a record of people with dyslexia who became well known in their fields. As the list grew longer and longer, I switched to keeping a record of just the fields. In medicine, individuals with dyslexia were likely to be found in radiology, where the ability to read patterns is central. In engineering and computer technology, they gravitated toward design and pattern recognition. In business, individuals with dyslexia, such as Paul Orfalea and Charles Schwab, tended to focus on high finance or money management, where forecasting trends and making inferences from large patterns of data are critical. My brother-in-law, an architect, told me that his former firm never allowed letters from its architects to go out without two spell-checks. Artists with dyslexia include sculptors such as Rodin and the painters Andy Warhol and Picasso. Actors include Danny Glover, Keira Knightley, Whoopi Goldberg, Patrick Dempsey, and Johnny Depp.

Two other examples come closer to home. When I was pregnant, I was referred to a world-famous radiologist in Boston for an ultrasound. As I lay waiting my turn, I overheard technicians talking about how people from around the world flew to this radiologist's clinic because she was the best. My antennae went up. As unobtrusively as possible, I asked them what made her great, and they instantly replied that it was her unerring ability to find unrecognized patterns within seconds. Later I learned that she and her father have a family history of dyslexia.

I had a similar experience during a recent trip to Barcelona. For five days I walked through the streets mesmerized by the brilliant designs, whimsical creations, and outrageous use of color in the churches and buildings designed by the great Spanish architect Antonio Gaudi. I became convinced that Gaudi had been dyslexic. Bingo. Every biography of Gaudi recounts the terrible time he had learning and reading as a child. He barely made it out of school, but once he did, he went on to become one of the outstanding fin de siècle Spanish artists and the patron architect of Barcelona.

How can we explain the preponderance of creativity and "thinking outside the box" in many people with dyslexia? As my son Ben asked, is the brain of a person with dyslexia forced to use the right hemisphere because of problems in the left hemisphere, thereby strengthening all the right-hemispheric connections and developing sometimes unique strategies for doing all kinds of things? Or are the right-hemisphere connections more dominant and creative from the start, therefore taking over activities such as reading? The neurologist Al Galaburda suspected that both scenarios might be correct in part: "Initially, circuits of the left-hemisphere type that do not form allow right-hemisphere circuits to populate empty synapses. Later, since they do not read, they get better at other things, especially since they have a good machinery for it."

There are no definitive answers to the questions raised by the preliminary evidence, but multilevel approaches that integrate information regarding behavior, cognition, neurological structures, and the genetics of dyslexia provide a good place to begin. The genetic foundation is pivotal. There are no "reading"-specific genes per se, but this doesn't mean that there are no genes linked to weaknesses in some of the older regions, which form the reading brain, and which are potentially linked to strengths in others. One future direction in dyslexia research will be to connect our knowledge about behavioral strengths and structural weaknesses to genetic information to see whether from the very outset some children with dyslexia have right hemispheres poised for building cathedrals.

Over eighty years ago, Samuel Orton first presented his provocative hypothesis about the failure of the brain's two hemispheres to integrate their stored images. More than fifty years later, Norman Geschwind wrote a paper titled simply, "Why Orton was Right." Geschwind listed thirteen conclusions about dyslexia that he and Orton shared and that should be incorporated into any explanation of dyslexia. Beginning with the genetic basis of dyslexia and possible structural differences in brain organization, this list went on to include remarkable spatial talents found in affected family members and also in some unaffected relatives; an unexpected ability to read equally well upside down or in a mirror (as my son and Leonardo da Vinci are known to have done); other unusual traits like dysgraphia; unusual speech, affect, and motor findings that aren't expressed in every case, but that need more intensive exploration (such as stuttering, ambi-dexterity, clumsiness, and emotional issues); and slowness in the acquisition and development of speech and language systems.

Geschwind's discussion of why Orton was right provides a checklist of what remains for twenty-first-century researchers to address before the puzzle of dyslexia can be satisfactorily explained. Using the example of a disease, sickle-cell anemia (whose gene simultaneously protects against malaria), Geschwind went on to make some observations that are as astute today as they were then:

> Dyslexics themselves are frequently endowed with high talents in many areas. . . . I would suggest to you that this is no accident. If certain changes on the left side of the brain lead to superiority of other regions, particularly on the right side of the brain, then there would be little disadvantage to the carrier of such changes in an illiterate society; their talents would make them highly successful citizens. . . . We are, thus, led to the paradoxical notion that the very same anomalies on the left side of the brain that have led to the disability of dyslexia in certain literate societies also determine superiority in the same brains.

These observations, like most of Geschwind's legendary ideas, were forerunners of empirical research in dyslexia that is only now catching up to him. Geschwind's early death prevented him from seeing how many of his insights continue to shape the field through his own direct contributions, through the work of his students, and through a dyslexia research program that began with him and that continues today to connect behavior to structure to neurons, and ultimately to genes.

The program of research envisioned by Geschwind began more than two decades ago with a chance discovery at Boston City Hospital: the carefully preserved brain of a person with dyslexia. No one knew what to do with it, so the brain was given to Geschwind, who knew exactly what to do. He promptly turned it over to two of his young students in neurology, Al Galaburda and Thomas Kemper, who proceeded to make careful studies, first of the macrostructure of several anatomical areas in this brain and then of the microstructure of regions important to reading.

Not long after this, another significant event occurred. Geschwind and Galaburda, along with the Orton Dyslexia Society, set up a brain bank, which over time became the repository of a few preserved brains of individuals with dyslexia at Beth Israel Hospital. This led to a discovery that continues to have implications for the current findings from right-hemisphere imaging. In most people the planum temporale (PT)—a triangular area on the temporal lobe that is involved in language and includes part of Wernicke's area—is larger in the left hemisphere than in the right. Galaburda and Kemper found that this asymmetry was not present in the brains of adults with dyslexia; rather, the two hemispheres were symmetric, because the right-hemisphere PT was larger than usual.

Galaburda and his team took these findings as an indication that lateralization is not completed or not the same in dyslexia—an interpretation that has implications for the development of many language processes. They suggested that the atypically large right-hemisphere planum temporale might result from a reduction of the natural pruning of cells that occurs during prenatal development. This could lead to increased numbers of PT neurons that then form new connections in the right hemisphere and a whole new cortical architecture in dyslexia. The potential importance of this explanation lost ground when attempts to find similar symmetries in living dyslexic persons by fMRI had mixed outcomes.

The inconclusiveness at the structural level prompted investigations at the cellular level. Using painstaking cytoarchitectonic methods, Galaburda and his colleagues studied the microstructure, number, and neuronal migration patterns of cells found in areas suspected to be aberrant in dyslexia. They found ectopic cells that had migrated during early prenatal development in several areas related to language and reading: the left planum temporale, several thalamic areas, and the visual cortex regions. Changes in neuronal migration in any of these areas could affect precise and efficient neuronal communication in regions that make up parts of the reading circuit.

For example, Galaburda's research team found that the magnocellular system—cells that are responsible for fast or transient processing—appeared consistently aberrant in at least two centers that are critical for reading within the thalamus, the brain's internal switchboard: the lateral geniculate nucleus (LGN), which helps coordinate visual processing; and the medial geniculate nucleus (MGN), which helps coordinate auditory processing. Once again, differences were found between the hemispheres, with the right hemisphere having more large neurons than the left. Galaburda argued that these cellular differences could affect the speed of information needed to process written language and might indicate that a different reading circuit is used in dyslexia.

As Galaburda cautiously noted, we do not yet know whether any of these differences are the source or a consequence of reading failure. What is emerging is that various neuronal changes, if found in important regions (i.e., the older structures necessary for reading), could disrupt the neuronal efficiency necessary for reading, thus promoting the formation of an altogether different reading circuit. Such a perspective would bring together many of the historical hypotheses about dyslexia based on deficits in structures, speed of processing, and changed circuitry.

Two unusual types of research illuminate this conclusion. One of them involves testing the effects of neuronal-level dysfunction in genetically selected mice, sometimes called, with tongue in cheek, the supermice. When the neuroscientist Glenn Rosen at Beth Israel induced a small lesion in the auditory cortex of these mice, neuronal anomalies formed in the thalamus, similar to those found earlier in the dyslexic brains. Most important, as a result of the lesions, the mice could no longer process rapidly presented auditory information. In other words, Glenn's animal model shows how wayward cells in important regions can cause problems in processing information efficiently.

A study by neurologists in Boston illustrates a similar set of principles for humans who have a rare genetic seizure disorder, periventricular nodular heterotopia. In this disorder, rogue cells form nodes in odd places next to the ventricles of the brain before birth. These nodes are analogous to the lesions induced in the supermice: they shouldn't be there, and at a certain point they become disruptive. In this case, the nodes caused seizures later in life—and also something else.

One of the authors of this study, Dr. Chang, came to me and to my colleague Tami Katzir perplexed by one behavioral trait found in all the patients: very poor reading fluency. Some patients had childhood diagnoses of dyslexia; some did not. Some had phonological weaknesses; some did not. But all were unexpectedly slow readers. Tami and I realized that these patients provided unexpected evidence of the many sources of problems with fluency, whether in adults or in children with reading disabilities.

These studies collectively illustrate several important principles. They show how different the pathways to inefficiency and impaired reading fluency can be, and how varied the sources of developmental dyslexia can be. The patients with seizures indicate that reading failure can result from dysfunction in multiple regions: for example, there were nodes in areas that might have affected visual efficiency, and other nodes in areas that might have affected weakened phonological processing. Both led to inefficient reading. What these cases don't explain is a reason for overreliance on the right hemisphere in some cases of dyslexia, but they do show how a wide variety of left-hemisphere weaknesses could force the brain to use analogous right-hemisphere areas.

. . . .

ONE HYPOTHESIS THAT emerges from this work follows Geschwind's logic. The genes that form the basis for a strengthened right hemisphere could have been highly productive in preliterate societies, but when these same genes are expressed within a literate society, they put structures in the right hemisphere in charge of the precise, time-based functions of reading. These functions then would be performed in the unique ways of the right hemisphere, rather than in the more precise, time-efficient ways of the left hemisphere. In the case of reading, that situation would lead inevitably to difficulties.

As an eminent geneticist has observed, reading is influenced by a number of genes whose presence may increase the risk of reading problems, but which do not cause such problems in the way that a single gene can cause a certain disease. For example, in the disease cystic fibrosis, only one gene determines the phenotype, or genetic outcome. By contrast, reading is based on many older processes and is therefore so complex that no one gene would likely ever determine all forms of breakdown in reading. In other words, there will be more than one phenotype.

The geneticist Elena Grigorenko of Yale underscores this point. After conducting a sweeping analysis of studies on the genetic regions associated with dyslexia, she concluded that the studies indicate multiple loci, not single genes. This conclusion makes a great deal of sense in light of the emerging subtypes of readers. As observed by Bruce Pennington and the Colorado research group, subtypes—such as readers with phonological deficit, fluency deficit, "double deficit," and orthographic deficit—may ultimately prove to be the behavioral manifestations of several phenotypes. And because of the different demands of various written languages, some phenotypes may be more prevalent in regular orthographies like German, whereas others may be more prevalent in less transparent languages like English, or in different systems like Chinese and Japanese logosyllabaries.

The idea that there are genetic differences in dyslexia in other languages has received preliminary support from some international research. Finnish and Swedish researchers presented data on one genetic location—called DCDC2—found on chromosome 6 that characterized many persons with dyslexia in German, with its prevalence of fluency deficits. For English-speakers, researchers at Yale and Colorado found data to support this location, but for only 17 percent of their subjects with dyslexia. Intriguingly, we find in our research on subtypes that about 17 percent of our subjects have only fluency-related deficits.

There is a fascinating twist to DCDC 2's story that relates

back to the notion of a different reading circuit in dyslexia. Using an animal model, Yale researchers found that when this genetic locus is not allowed to be expressed, young neurons do not migrate to the right-hemisphere cortex. These researchers hypothesized that similar genetic variations in children with dyslexia could lead to the formation and use of "less efficient circuits for reading."

In a different study, a large Finnish family with a long genetic history of dyslexia showed genetic variations in an area called ROBO1. Fascinatingly, in light of Orton's earlier hypotheses, ROBO1 helps "shape neural connections between the two sides of the brain during development and may be impaired in dyslexia." Also, in these studies, two distinct areas appear in two regular languages—a fact that buttresses multidimensional explanations for dyslexia and the work on subtypes within a single language.

Other support comes from one of the largest and most established genetic programs in the United States, the Colorado Twin Study, in which psychologist Dick Olson and other researchers followed over 300 pairs of dizygotic (fraternal) and monozygotic (identical) twins in kindergarten and later. This group found that children's abilities in reading, phoneme awareness, and rapid naming (RAN) showed substantial genetic effects and some environmental effects. Most important for understanding possible subtypes in dyslexia, phonological skills and rapid naming each showed separate, significant heritability.

If these results are replicated, it could mean that there are separate genes at work for the two sets of processes known to characterize well-documented subtypes of reading disabilities in English, and to predict dyslexia in many languages. If future studies pinpoint the different phenotypes and their structural and behavioral characteristics, deficits, and strengths, we could supply many of the puzzle pieces still missing from the history of dyslexia.

And if there are several phenotypes, some children might inherit dyslexia from both sides of the family. If I think about the genetic history of subtle or blatant reading disabilities in my own son Ben's family tree, he and his brother David look like clear examples of what Orton and Geschwind observed. Despite the fact that David is a gifted writer, avid soccer player, and is supposedly unaffected, his problems with word retrieval and dysgraphia have defied all efforts at remediation. David's profile and Ben's double deficits could derive from a combination of genes from each side of the family. My husband's father, Ernst Noam, was a European intellectual trained in German law, but he was never able to practice law in Hitler's Germany. My husband's sister is convinced, from her father's unusual learning history, that he had some form of reading disability, even though he read in four languages. My own maternal great-great-grandfather reversed numbers and letters so pronouncedly that the fact was deemed noteworthy in his description in a history of the state of Indiana. Gil's and my siblings, cousins, and nieces and nephews on both sides are an assortment of successful artists, engineers, lawyers, businessmen, and surgeons, several of whom have dealt with subtle and not-so-subtle learning issues.

Geschwind wrote at some length about the need to understand genetically all that is going on below the surface of our understanding of "unaffected" relatives. He noted, for example, Orton's own "remarkable spatial talents." I didn't need to go very far to see David's dysgraphia and word-retrieval issues, but I never examined my own learning history until I sat down to write this chapter. On the surface my reading process is unremarkable, but my word-retrieval processes require no small effort—invisible only because my great love of words provides me with ready alternatives.

And there is something else that I've never connected until now. Years ago my secret fantasy was to become a pianist. It was a very brief fantasy, dashed when my otherwise gentle instructor told me she always loved listening to me play Mozart, Chopin, or Beethoven, but it was never what the composer intended. She said I had my own timing, which always differed from the composer's, and she didn't think that this could ever change. In a flash I knew why all those poor kids I had accompanied on the piano in their recitals had always sounded a little off tempo. The problem had been my timing, not theirs! Only now do I think that my unusual time patterns in reading music notation may be a manifestation of my own genetically based differences in speed of processing. When a child has dyslexia, there are no "unaffected" relatives in the family. We are all affected, every day, as anyone who has a child, grandchild, or sibling with dyslexia knows. But we may be affected in more ways than we realize—ways that can open the door to understanding many idiosyncrasies that make all of us in the genetic family of dyslexia such a richly diverse group.

I am somehow less interested in the weight and convolutions of Einstein's brain than in the near certainty that people of equal talent have lived and died in cotton fields and sweatshops.
—STEPHEN JAY GOULD

Finally, the single most important implication of research in dyslexia is not ensuring that we don't derail the development of a future Leonardo or Edison; it is making sure that we do not miss the potential of any child. Not all children with dyslexia have extraordinary talents, but every one of them has a unique potential that all too often goes unrealized because we don't know how to tap it.

We who work with these children seek to find methods that can realize their potential. After all is said and done, the research on dyslexia from behavior to gene ultimately needs to connect what we know to what and how we teach and whether it works or doesn't work for a particular child. For reasons we've explored, children struggling to read aren't going to be helped by the one-size-fits-all approach that is typical in so many schools. Rather, we need teachers who are trained to use a toolbox of principles that they can apply to different types of children. And we need educational research that, as the policy maker Reid Lyon has often said, seeks to investigate and understand what emphases work best under what conditions for which children. There are no universally effective programs, but there are knowable principles that need to be incorporated in all programs about how we teach written language.

Some of the most important principles are as old as written language itself. For years my coworkers and I at the Center for Reading and Language Research have used our knowledge of what the brain does when it reads a word or a story to design and evaluate an intervention program (RAVE-O) that can address many linguistic weaknesses of struggling readers. We never realized we were reinventing a program with some of the same principles used in the first known reading pedagogy—that of the Sumerians. We may package our teaching in wholly different ways, but like the Sumerians we give daily emphasis to each of the major linguistic and cognitive processes used by the brain to read: semantic families of words to teach semantic depth and to facilitate retrieval; awareness of sounds within words and their connections to letter representation; automatic learning of orthographic letter patterns; syntactic knowledge; and morphological knowledge. Unlike the Sumerians, we also use multiple strategies for fluency and comprehension. Like the Sumerians we want every struggling reader to know as much about a word as possible; perhaps unlike them, we want each child to have fun learning.

Those of us who work with children want them to realize that although they may learn differently, each one of them can and will learn to read. It is our job, not theirs, to find out how best to teach them. A decade of research on various interventions with my colleagues Robin Morris and Maureen Lovett supports efforts to do just that.

Future efforts at our lab and at centers around the country are now linking intervention not only to behavioral changes in response to intervention but also to neuronal changes. For example, we are working with John Gabrieli's group at MIT to see whether important areas in the brain change in readers with dyslexia before and after our program is taught to them. Good teachers don't need neuroscience to know that multiple aspects of oral and written language are important, but educational research informed by neuroscience can identify what works best for an individual child.

It can do so by allowing us to observe which structural regions in the brains of children are engaged during particular tasks, and how these may or may not change after a specific set of emphases in treatment.

. . . .

THESE NEW DIRECTIONS are changing the way I think about dyslexia as a researcher, and as a parent. If some version of the emerging theories about reliance on the right hemisphere in dyslexia turns out to be true for some children, or for many, this could open up relatively unexplored avenues for teaching the differently organized brain, with its unique mix of strengths and challenges. Finally, all this research on children who learn to read in different ways becomes part of the great body of knowledge about how all of us learn to read. Regardless of its ultimate interpretation over time, this area of research insists that we go beyond what we have learned over the past two decades into new barely explored territories. Appropriately enough, going beyond what we know is, in fact, the last undertaking of this book.

Chapter 9

CONCLUSIONS: FROM
THE READING BRAIN TO
"WHAT COMES NEXT"

Each torpid turn of the world has such disinherited children to
whom neither what's been nor what is to come, belongs,
For what comes next *is too large and remote for humankind.*
—RAINER MARIA RILKE

Reading is an act of interiority, pure and simple. Its object is
not the mere consumption of information. . . . Rather, reading
is the occasion of the encounter with the self. . . . The book is
the best thing human beings have done yet.
—JAMES CARROLL

In the clash between the conventions of the book and the
protocols of the screen, the screen will prevail. On this screen,
now visible to one billion people on earth, the technology of
search will transform isolated books into the universal library
of all human knowledge.
—KEVIN KELLY

EVERY SOCIETY WORRIES OVER THE FUTURE OF ITS
young and the challenges they will face. No one describes
the accelerating pace of those challenges at this moment in human evolution more compellingly than the futurist and inventor

Ray Kurzweil. His visionary work depicts the staggering shifts that may occur as the 100 trillion neural connections in our brains extend exponentially through the technological, nonbiological intelligence we have invented:

> We can have confidence that we will have the data-gathering and computational tools needed by the 2020s to model and simulate the entire brain, which will make it possible to combine the principles of operation of human intelligence with the forms of intelligent information processing.—We will also benefit from the inherent strength of machines in storing, retrieving, and quickly sharing massive amounts of information. We will then be in a position to implement these powerful hybrid systems on computational platforms that greatly exceed the capabilities of the human brain's relatively fixed architecture. . . .
>
> How can we, limited by our current brain's capacity for 10^{16} to 10^{19} calculations per second, even begin to imagine what our future civilization in 2099—with brains capable of 10^{60} calculations per second—will be capable of thinking and doing?

One thing we can imagine is that our capacities for good and for destruction will also be exponentially increased. If we are to prepare for such a future, our ability to make profound choices must be honed with a rigor rarely practiced by learners in past generations. If the species is to progress in the fullest sense, such preparations require singular capacities for attention and decision making that incorporate a desire for the common good. In other words, to prepare for what comes next demands the absolute best of what we possess in the present adaptation of the reading brain, as it already begins to undergo its next generation of changes.

I differ with Kurzweil's implicit assumption that an exponential acceleration of thought processes is altogether positive. In music, in poetry, and in life, the rest, the pause, the slow movements are essential to comprehending the whole. Indeed, in our brain there are "delay neurons" whose sole function is to slow neuronal transmission by other neurons for mere milliseconds. These are the inestimable milliseconds that allow sequence and order in our apprehension of reality, and that enable us to plan and synchronize soccer moves and symphonic movements.

The assumption that "more" and "faster" are necessarily better requires vigorous questioning, especially since this assumption already increasingly influences everything in American society, including how we eat and how we learn, with doubtful benefits. For example, will the accelerated rate of change already experienced by our children have consequences that radically affect the quality of attention that can transform a word into a thought and a thought into a world of unimagined possibility? Will this next generation's capacity to find insights, pleasure, pain, and wisdom in oral and written language be dramatically altered? Will their relationship to language be fundamentally changed? Will the present generation become so accustomed to immediate access to on-screen information that the range of attentional, inferential, and reflective capacities in the present reading brain will become less developed? And what of future generations? Are Socrates' concerns about unguided access to information more warranted today than they were in ancient Greece?

Or will the demands of our new information technologies—to multitask, and to integrate and prioritize vast amounts of information—help to develop equally if not more valuable new skills that will increase our human intellectual capacities, our quality of life, and our collective wisdom as a species? Could the acceleration of such intelligence allow us more time for reflection and for the pursuit of the good for humanity? If so, will this next set of intellectual skills produce a new disenfranchised group of differently wired children equivalent to the dyslexic readers of the present? Or will we now be more prepared to view children's learning differences in terms of different patterns of brain organization, with genetic variations that bestow both strengths and weaknesses?

Dyslexia is our best, most visible evidence that the brain was never wired to read. I look at dyslexia as a daily evolutionary reminder that very different organizations of the brain are possible. Some organizations may not work well for reading, yet are critical for the creation of buildings and art and the recognition of patterns—whether on ancient battlefields or in biopsy slides. Some of these variations of the brain's organization may lend themselves to the requirements of modes of communication just on the horizon.

In the twenty-first century we are poised to change significantly and rapidly in ways that most of us can barely predict or fully comprehend. It is within this pronounced sense of transition that I locate this book's central themes about the evolution, development, and different organizations of the reading brain. The evolution of writing and the development of the reading brain give us a remarkable lens on ourselves as a species, as the creators of many oral and written language cultures and as individual learners with different and expanding forms of intelligence.

In this final chapter I use the lens of reading to look back over several major insights, and then to venture "beyond the text." There, in that uncharted territory, I want to consider the implications of this information for the present generation of children and for the next. And by the end, I want to reflect on what we should strive with all our power to preserve in the reading brain, before the transition to its next rearrangement is complete.

Reflections on Reading's Evolution

My overarching reaction to the evolution of the reading brain is surprise. How could a tiny set of token symbols flower in such a relatively short time into a full-blown writing system? How could a single cultural invention less than 6,000 years old change the ways the brain is connected within itself and the intellectual possibilities of our species? And then there's a deeper surprise: how miraculous it is that the brain can go beyond itself, enlarging both its functions and our intellectual capacities in the process. Reading illuminates how the brain learns new skills and adds to its intelligence: it rearranges the circuits and connections among older structures; it capitalizes on the ability to commit areas to specialization, particularly pattern recognition; and it illustrates how new circuits can become so automatic that more cortical time and space can be allocated to other, more complex, thought processes. In other words, reading displays how the most basic design principles in the brain's organization underlie and shape our continuously evolving cognitive development.

The brain's design made reading possible, and reading's design changed the brain in multiple, critical, still evolving ways. The reciprocal dynamics shine through the birth of writing in the species and through the acquisition of reading in the child. Learning to read released the species from many of the former limitations of human memory. Suddenly our ancestors could access knowledge that would no longer need to be repeated over and over again, and that could expand greatly as a result. Literacy made it unnecessary to reinvent the wheel and thus made possible the more sophisticated inventions that would follow, like a machine that can read to those who can't, invented by Ray Kurzweil.

Simultaneously, the capacity of literacy for rapid-fire performance released the individual reader not only from the restrictions of memory but from those of time. By its ability to become virtually automatic, literacy allowed the individual reader to give less time to initial decoding processes and to allocate more cognitive time and ultimately more cortical space to the deeper analysis of recorded thought. Developmental differences in the circuit systems between a beginning, decoding brain and a fully automatic, comprehending brain span the length and breadth of the brain's two hemispheres. A system that can become streamlined through specialization and automaticity has more time to think. This is the miraculous gift of the reading brain.

Few inventions ever did more to prepare the brain and poise the species for its own advancement. As literacy became widespread in a culture, the act of reading silently invited each reader to go beyond the text; in so doing, it further propelled the intellectual development of the individual reader and the culture. This is the biologically given, intellectually learned generativity of reading that is the immeasurable yield of the brain's gift of time.

The biological evidence for this view begins with the realization that structurally there is little to differentiate our brain today from that of nonliterate humans 40,000 years ago. We share our brain structures with our Sumerian and Egyptian ancestors. How we use and connect these structures, however, creates a distinction, as the comparative reading of different writing systems like hieroglyphs and alphabets illustrates. The pioneering work of Charles Perfetti, Li-Hai Tan, and their group demonstrates that each writing system—ancient or new—uses many similar and some unique structural connections. A brain wired to read Egyptian hieroglyphs or Chinese characters activates some areas never used to read the Greek or English alphabet, and vice versa. The variety of these adaptations is fresh evidence of the brain's innate potential for rearranging itself to perform new functions.

With the birth of writing systems, changes occurred in more than just the brain's circuitry. As the classicist Eric Havelock asserts, the Greek alphabet represents a psychological and pedagogical revolution in human history: the process of writing released an unprecedented ability to achieve novel thoughts. Some of our finest cognitive neuroscientists study the neurological basis for this new ability in all comprehensive writing systems, not only alphabets. They describe how the reordering of the brain's basic computations that occurs during the acquisition of reading becomes the neuronal basis for new thoughts. In other words, the new circuits and pathways that the brain fashions in order to read become the foundation for being able to think in different, innovative ways.

The reading revolution, therefore, was both neuronally and culturally based, and it began with the emergence of the first comprehensive writing systems, not the first alphabet. The increased efficiency of writing and the memory it freed contributed to new forms of thought, and so did the neuronal systems set up to read. New thought came more readily to a brain that had already learned how to rearrange itself to read; the increasingly sophisticated intellectual skills promoted by reading and writing added to our intellectual repertoire, and continue to add to it.

To come to this understanding, we must reflect on a question: what are the skills promoted by literacy that are not found in oral cultures? With the creation of the earliest token symbols came the first known accounting system, and with it the enhanced decision making that occurs when more and better information becomes available. It would appear, therefore, that the first known symbols (other than cave drawings) were in the service of economy—and economics. With the first comprehensive writing systems—Sumerian cuneiform and Egyptian hieroglyphs—simple accounting became systematic documentation, which led to organizational systems and codification, which in turn facilitated significant intellectual advances. By the second millennium BCE, Akkadian literary works had begun to classify the entire known world, as exemplified by the encyclopedic *All Things Known in the Universe*, the legal masterpiece *Code of Hammurabi*, and various remarkable medical texts. The scientific method itself had its origins in our ancestors' growing ability to document, codify, and classify.

An increasing linguistic awareness is evident in many places, beginning with the Sumerian methods for teaching reading. The methods they used in their *e-dubba* ("tablet house") contributed to a heightened understanding of the different properties of words: the multiple semantic or meaning relationships among words; their different grammatical functions; the combinatorial capacities within words that allow new words to be formed from existing stems and morphemes; and the different pronunciations across dialects and languages.

The young Sumerians' task of painstakingly copying lists of written words on the other side of the teacher's tablet gave students time to reflect on the words they were inscribing. This contributed not only to the gradual development of linguistic awareness, but also to the process of deliberation itself. Centuries later, Akkadian works like *Gilgamesh, Dialogue on Pessimism*, and many preserved Ugaritic documents helped make visible the feelings, thoughts, trials, and joys of these grown pupils, and revealed their inner lives.

These ancient works became timeless witnesses to the emergence of what we often think of as modern consciousness.

Few scholars are more eloquent about the contributions of literacy to the emergence of consciousness in the ancient world than the Jesuit cultural historian Walter Ong. In his lifelong study of the relationship between the spoken word and literacy, Ong reframed the question of the unique contributions of reading in a way that may help us understand our own current transition to more digital modes of communicating. Two decades ago, Ong asserted that the real issue in human intellectual evolution is not the set of skills advanced by one cultural mode of communication versus another, but the transformative changes bestowed on humans steeped in both. In a prescient passage, Ong wrote:

> The interaction between the orality that all human beings are born into and the technology of writing, which no one is born into, touches the depths of the psyche. It is the oral word that first illuminates consciousness with articulate language, that first divides subject and predicate and then relates them to one another, and that ties human beings to one another in society. Writing introduces division and alienation, but a higher unity as well. It intensifies the sense of self and fosters more conscious interaction between persons. Writing is consciousness-raising.

To Ong, new understandings of human consciousness were the real changes rendered when oral and written language converged: reading changed how human beings could think about thinking. From Levin's disclosures in *Anna Karenina* to a spider's predicament in *Charlotte's Web*, the ability to see another's thoughts makes us doubly aware—of the other's consciousness and of our own. Through our ability to study people's thought processes across 3,000 years, we are able to internalize the consciousness of human beings we could never otherwise imagine, including that of the greatest apologist of oral traditions, Socrates. It is only because we can read the product of Plato's ambivalence that we can come to understand Socrates and the universal nature of his concerns.

When all is said and done, of course, Socrates' worries were not so much about literacy as about what might happen to knowledge if the young had unguided, uncritical access to information. For Socrates, the search for real knowledge did not revolve around information. Rather, it was about finding the essence and purpose of life. Such a search required a lifelong commitment to developing the deepest critical and analytical skills, and to internalizing personal knowledge through the prodigious use of memory, and long effort. Only these conditions assured Socrates that a student was capable of moving from exploring knowledge in dialogue with a teacher to a path of principles that lead to action, virtue, and ultimately to a "friendship with his god." Socrates saw knowledge as a force for the higher good; anything—such as literacy— that might endanger it was anathema.

Socrates' concerns might have been partly addressed through a more nuanced understanding of how inextricably related knowledge and literacy are, and how important they are to the development of the young. Ironically, today's hypertext and online text provide a dimension of virtual dialogue to reading in computer-based presentations. The contemporary scholar John McEneaney argues that the "dynamic agency of on-line literacy challenges the traditional roles of reader and author, as well as the authority of text." Such reading requires new cognitive skills that neither Socrates nor modern educators totally understand. We are only at the beginning of analyzing the cognitive implications of using, for instance, the browser "back" button, URL syntax, "cookies," and "pedagogical tags" for enhancing comprehension and memory. These tools have extremely promising implications for the intellectual development of the users, particularly users with discrete areas of weakness that applied learning technologies can address directly and well. As the applied technology expert David Rose and his group persuasively demonstrate, digital texts can offer choice to teacher and learner: "choice in appearance, in level of support, in type of support, in method of response, in content . . . all key to engagement." And the engagement of our learners is as important today as it was in the Athenian court-yards.

There are deeper meanings in these Socratic concerns, how-ever. Throughout the story of humankind, from the Garden of Eden to the universal access provided by the Internet, questions of who should know what, when, and how remain unresolved. At a time when over a billion people have access to the most extensive expansion of information ever compiled, we need to turn our analytical skills to questions about a society's responsibility for the transmission of knowledge. Ultimately, the questions Socrates raised for Athenian youth apply equally to our own. Will unguided information lead to an illusion of knowledge, and thus curtail the more difficult, time-consuming, critical thought processes that lead to knowledge itself? Will the split-second immediacy of in-formation gained from a search engine and the sheer volume of what is available derail the slower, more deliberative processes that deepen our understanding of complex concepts, of another's inner thought processes, and of our own consciousness?

At the start of this book I quoted the technology expert Ed-ward Tenner, who asked whether our new information tech-nology would "threaten the very intellect that created it." This book's questions are not quixotic efforts to prevent the spread of technology—whose indisputable worth transforms all our lives. Tenner's concerns are the technological analogue both of Socrates' concerns and of the issues discussed below about what the read-ing brain contributes to the intellectual formation of the species and the child. The question that emerges, therefore, is this: what would be lost to us if we replaced the skills honed by the reading brain with those now being formed in our new generation of "digital natives," who sit and read transfixed before a screen?

The evolution of writing provided the cognitive platform for the emergence of tremendously important skills that make up the first chapters of our intellectual history: documentation, codifica-tion, classification, organization, interiorization of language, consciousness of self and others, and consciousness of conscious-ness itself. It is not that reading directly caused all these skills to flourish, but the secret gift of time to think that lies at the core of the reading brain's design was an unprecedented impetus for their growth. Examining the development of these skills through the "natural history of reading" shows in slow motion how far our species has come in the 6,000 years since literacy emerged, as well as what it stands to lose.

Reflections on the "Natural History" of Reading

Each brain of each ancestral reader had to learn to connect multiple regions in order to read symbolic characters. Each child today must do the same. Young novice readers around the globe must learn how to link up all the perceptual, cognitive, linguistic, and motor systems necessary to read. These systems, in turn, depend on utilizing older brain structures, whose specialized regions need to be adapted, pressed into service, and practiced until they are automatic.

For this to happen in the absence of any genetic transmission specific to reading requires explicit learning and explicit teaching, all in a relatively brief time. Despite the fact that it took our ancestors about 2,000 years to develop an alphabetic code, children are regularly expected to crack this code in about 2,000 days (that is, by six or seven years of age), or they will run afoul of the whole educational structure—teachers, principals, family, and peers. If reading is not acquired on society's schedule, these suddenly disinherited children will never feel the same about themselves. They will have learned they are different, and no one ever tells them that, evolutionarily, this might be for good reason.

As we recognize the neuronal high-wire act that the young brain has to accomplish to acquire reading, we as a society can begin to teach individual children. Some children need more help than others with one or more of the parts of reading. The more we learn about those parts, the better able we will be to teach all children. Within such a perspective there can be no one-size-fits-all instruction. Our expanding knowledge about the development of reading has the potential to contribute to two all-important goals: understanding the magnitude of the reading brain's accomplishments, and improving the opportunities for every individual child in the next generation to learn to read.

The developmental transformations that mark the way to reading expertise begin in infancy, not in school. The amount of time the child spends listening to parents and other loved ones read continues to be one of the best predictors of later reading. As they listen to stories of Babar, Toad, and Curious George and say "good night moon" every evening, children gradually learn that the mysterious notations on the page make words, words make stories, stories teach us all manner of things that make up the known universe.

Their world of stories, words, and magic letters is a microcosm of the thousands of words, concepts, and perceptions that go into the development of the young brain readying itself to read. The more young children are engaged in conversation, the more they will acquire words and concepts. The more young children are read to, the more they will understand the language of books and increase their vocabulary, their knowledge of grammar, and their awareness of the tiny but very important sounds inside words. The full sum of this tacit knowledge—the similar sounds in "hickory, dickory, dock"; the multiple meanings of "bear"; the fearful thoughts of Wilbur the pig—prepares the young child's brain to connect visual symbols to all that stored knowledge.

The development of reading, therefore, has two parts. First, the ideal acquisition of reading is based on the development of an amazing panoply of phonological, semantic, syntactic, morphological, pragmatic, conceptual, social, affective, articulatory, and motor systems, and the ability of these systems to become integrated and synchronized into increasingly fluent comprehension. Second, as reading develops, each of these abilities is facilitated further by this development. Knowing "what's in a word" helps you read it better; reading a word deepens your understanding of its place in the continuum of knowledge.

This is the dynamic relationship between the brain's contribution to reading and reading's contribution to the brain's cognitive capacities. Children's phonological systems help them to develop an awareness of the sounds inside a word; this awareness helps them learn letter-sound rules; those rules help them learn to read more easily. Then, as children begin to read more and more, they become exquisitely attuned to the phonemic aspects in words, which makes reading easier. Similarly, children whose semantic systems are well developed know the meanings of more words, so that they are able to decode already known words faster. This adds to their repertoire of written words, which fosters their oral vocabulary, which prepares them to read even more sophisticated stories—which increases their knowledge of grammar, morphology, and relationships among words. "The rich get richer and the poor poorer." These developmental-environmental dynamics form the basis for making the great transition from "learning to read" to real reading, or not.

Fluent, silent comprehension in the later phases of reading development would have symbolized for Socrates the most dangerous moment in literacy, because it makes the reader autonomous. It gives each new reader time to make predictions, to form new thoughts, to go beyond the text, and to become an independent learner. Imaging studies confirm that the fluent reading brain activates newly expanded cortical regions across frontal, parietal, and temporal lobes of both hemispheres during comprehension processes such as inference, analysis, and critical evaluation. These are some of the very intellectual skills Socrates feared would be lost if literacy was allowed to spread.

Other concerns of Socrates are less resolved during the developmental transition to "expert reading." First, do most young readers, in fact, really learn to use their imagination fully, or to use their independent, probing, analytical processes? Or are these more time-demanding skills increasingly derailed by the seemingly limitless information children now receive on-screen? Do young readers who spend a disproportionate amount of their reading time on-screen, as opposed to in the pages of a book, develop differently in their ability to identify with Jane Eyre, Atticus Finch, and Celie?

I do not question the extraordinary ways the digital world

brings to life the realities and the perspectives of other people and cultures. I do wonder whether typical young readers view the analysis of text and the search for deeper levels of meaning as more and more anachronistic because they are so accustomed to the immediacy and seeming comprehensiveness of the on-screen information—all of which is available without critical effort, and without any apparent need to go beyond the information provided. I ask, therefore, whether our children are learning the heart of the reading process: going beyond the text.

Recently I read an essay in the *Wall Street Journal*, headed "How Low Can They Go?" It was about the current decline in verbal SAT scores. The writer described recent changes in the SAT test that resulted in more emphasis on reading skills than on vocabulary, thereby rewarding students with more refined analytical skills and penalizing those less prepared to discern and evaluate the underlying meaning of a text. He observed that students of forty years ago probably would do better in this test format than today's students, who appear far less capable of reading critically. For this he blamed the schools, not the test.

Blame is rarely well distributed. The author of this essay may well be correct, but there are many reasons for a decline: some sociological, some political, and some cognitive. Many students who have cut their teeth on relatively effortless Internet access may not yet know how to think for themselves. Their sights are narrowed to what they see and hear quickly and easily, and they have too little reason to think outside our newest, most sophisticated boxes. These students are not illiterate, but they may never become true expert readers. During the phase in their reading development when critical skills are guided, modeled, practiced, and honed, they may have not been challenged to exploit the acme of the fully developed, reading brain: time to think for themselves.

Everyone involved in the education of the young—parents, teachers, scholars, policy makers—needs to ensure that each component of the reading process is sensibly, carefully, explicitly prepared for or taught from birth until full adulthood. Nothing, from knowledge about the word's smallest sounds in preschool to the ability to interpret T. S. Eliot's most subtle inferences in "Little Gidding," should be taken for granted along the way. And within children's particularly vulnerable transition to the level of fluent, comprehending reader we must exert our greatest efforts to ensure that immersion in digital resources does not stunt our children's capacity to evaluate, analyze, prioritize, and probe what lies beneath any form of information. We must teach our children to be "bitextual," or "multitextual," able to read and analyze texts flexibly in different ways, with more deliberate instruction at every stage of development on the inferential, demanding aspects of any text. Teaching children to uncover the invisible world that resides in written words needs to be both explicit and part of a dialogue between learner and teacher, if we are to promote the processes that lead to fully formed expert reading in our citizenry.

My major conclusion from an examination of the developing reader is a cautionary one. I fear that many of our children are in danger of becoming just what Socrates warned us against—a society of decoders of information, whose false sense of knowing distracts them from a deeper development of their intellectual potential. It does not need to be so, if we teach them well, a charge that is equally applicable to our children with dyslexia.

Reflections on Dyslexia and Thinking Outside the Box

In a book devoted to the reading brain it would be easy enough to skip over the contributions of a brain ill-suited to reading. But the squid who doesn't swim quickly has a lot to teach about how it learns to compensate. This is an imperfect analogy, to be sure, because the squid's ability to swim is genetic and a squid who can't swim quickly would very likely die. But *if* a poor-swimming squid not only didn't die, but went on to beget 5 to 10 percent of the squid population, one would have to ask what in the world that squid had going for itself that made it so successful despite the missing capacity. Reading isn't laid down genetically, and the child who can't learn to read doesn't die. More significantly, the genes associated with dyslexia have survived robustly.

The list of gifted figures with dyslexia—people such as Rodin and Charles Schwab—may be one reason why. Another reason is bound up in our human diversity. As Norman Geschwind often asserted, the diversity of our genetically endowed strengths and weaknesses allows us to form a society capable of meeting all our varied needs. Dyslexia, with its seemingly untidy mix of genetic talents and cultural weaknesses, exemplifies human diversity— with all the important gifts this diversity bestows on human culture. Picasso's *Guernica*, Rodin's *Thinker*, Gaudi's *La Pedrera*, and Leonardo's *Last Supper* are icons as real and as expressive of our intellectual evolution as any written text. That all these were created by individuals who more than likely were dyslexic is not coincidental.

The real tragedy of dyslexia is that no one tells this to the children who year after year publicly, humiliatingly, cannot learn to read, despite all their intelligence and despite the critical importance of just their type of intelligence for the species. Also, no one tells the children's peers. This view does not minimize the difficulties every child with dyslexia confronts in learning. On the contrary, it tells these children just how important they are to us all, and that it is up to us to find better ways of teaching this differently organized brain to learn to read.

One of the most hopeful applications of neuroscience concerns just that. The more we know about the development of the reading brain and the dyslexic brain, the better we are able, in our interventions, to target more specifically the particular parts or connections that are not developing in some children. Intervention in dyslexia—just as in reading that is developing typically— must explicitly address every component system of reading intensively and imaginatively, until some level of automaticity and comprehension is attained. This is a far more difficult and demanding task for a brain that is wired less efficiently for many written language processes, and that may well represent a different adaptation of the brain for reading.

It is in the highest interests of our society to protect the potential contributions of our children with dyslexia. As described in the work of Harvard scholar Gil Noam, there is a necessity that we help them endure what is difficult and foster their resilience, so that they are prepared to invent the next lightbulb when they are ready. I do not want to dwell on the waste that has been caused by years of ignorance about dyslexia and many other forms of learning disabilities. It is a sad chapter in the larger story that began when some of us learned to read, while others among us continued to build, create wondrous things, and think differently from the rest. Fortunately, the stories of the reading brain and the dyslexic brain are emerging as twinned tales in the larger saga of the great human family.

An appreciation of the genetic diversity that drives all these differences in our intellectual traits and skills is especially important during our transition to the near future. Not unlike Plato's ambivalence, this book is written from two perspectives—that of a passionate apologist for the reading brain's contributions to our intellectual repertoire, and that of a contributing participant in and vigilant observer of the technological changes which will help shape the next rearranged brains. Humans today do not need to be binary thinkers, and future generations certainly don't. As an apt Viennese expression puts it, "If two choices appear before you, there's usually a third."

In the transmission of knowledge the children and teachers of the future should not be faced with a choice between books and screens, between newspapers and capsuled versions of the news on the Internet, or between print and other media. Our transition generation has an opportunity, if we seize it, to pause and use our most reflective capacities, to use everything at our disposal to prepare for the formation of what will come next. The analytical, inferential, perspective-taking, reading brain with all its capacity for human consciousness, and the nimble, multifunctional, multimodal, information-integrative capacities of a digital mind-set do not need to inhabit exclusive realms. Many of our children learn to code-switch between two or more oral languages, and we can teach them also to switch between different presentations of written language and different modes of analysis. Perhaps, like the memorable image captured in 600 BCE of a Sumerian scribe patiently transcribing cuneiform beside an Akkadian scribe, we will be able to preserve the capacities of two systems and appreciate why both are precious.

. . . .

IN SUM, THE NATURAL history of reading development presents an exceedingly hopeful, but also cautionary tale about reaching the highest and deepest levels of reading. It's a magnificent, sometimes poignant, often humbling story that began thousands of years ago, in cultures that are known to us only because some human ancestors had the daring and the neuronal adaptability to preserve their debts and their yearnings on tablets of clay and rolls of papyrus.

Equally courageous, Socrates feared above all else that the "semblance of truth," conveyed by the seeming permanence of this written language, would lead to the end of the search for true knowledge, and that this loss would mean the death of human virtue as we know it. Socrates never knew the secret at the heart of reading: the time it frees for the brain to have thoughts deeper than those that came before. Proust knew this secret, and we do. The mysterious, invisible gift of *time to think beyond* is the reading brain's greatest achievement; these built-in milliseconds form the basis of our ability to propel knowledge, to ponder virtue, and to articulate what was once inexpressible—which, when expressed, builds the next platform from which we dive below or soar above.

To the Reader: A Final Thought

A book about how our species learned to leap beyond the text shouldn't have a last sentence. Gentle readers, it is all yours. . . .

ACKNOWLEDGMENTS

I T TOOK SEVEN YEARS AND 100 FRIENDS AND colleagues to complete this book. During this time more than twelve wonderful children (they are still coming) were born to members connected to the Center for Reading and Language Research, to our great joy. At the same time, we lost eight friends, all of whom contributed to the work in this book in very different ways: David Swinney, eminent cognitive scientist and lifelong friend; Michael Pressley and Steve Stahl, dedicated psychologists, humanists, and educators; Jane Johnson, tireless, beautiful advocate for persons with learning disabilities; Rebecca Sandak, young gifted neuroscientist; Merryl Pisha, one of the best of all reading teachers in Boston; Harold Goodglass, one of the finest neuropsychologists of the twentieth century; and Ken Sokoloff, my brilliant economist and beloved friend. I wish to mark each of their contributions to their respective fields and to me.

My personal gratitude begins at the Center for Reading and Language Research, which I have directed over the last decade at Tufts University. The Center is an evolving home to a group of unbelievably committed colleagues who have taught, tutored, and tested over 1,000 children and who have conducted research into everything from intervention for children with dyslexia to brain imaging of letter naming. It is the finest group of people with whom I have ever worked. At various times they have included: Katherine Donnelly Adams, Maya Alivasatos, Mirit Barzillai, Surina Basho, Terry Joffe Benaryeh, Alexis Berry, Kathleen Biddle, Kim Boglarksi, Ellen Boiselle, Joanna Christodoulou, Colleen Cunningham, Terry Deeney, Caroline Donelan, Wendy Galante, Yvonne Gil, Stephanie Gottwald (research coordinator), Alana Harrison, Jane Hill, Julie Jeffery, Manon Jones, Tami Katzir, Rebecca Kennedy, Anne Knight, Kirsten Kortz, Cynthia Krug, Jill Ludmar, Emily McNamara, Larina Mehta, Maya Misra, Lynne Tomer Miller (assistant director), Kiran Montague, Cathy Moritz, Elizabeth Norton, Beth O'Brien, Alyssa O'Rourke, Margaret Pierce, Connie Scanlon, Erika Simmons, Catherine Stoodley, Laura Vanderberg, and Kim Walls.

Honorary members of the Center include Pat Bowers and Zvia Breznitz, who spent sabbaticals with us, and Ginger Berninger: they are friends for life. Many people in the Center helped do "things" connected to this book for which I am forever grateful. Permissions: Pascale Boucicaut and Andrea Marquant. References: Kirsten Kortz (who deserves wings) and Katherine Donnelly Adams. Proofreading: my seminar students, Mirit Barzillai, Cathy Moritz, and Elizabeth Norton. Conceptual insights: my former student and now esteemed colleague in Haifa, Tami Katzir. Technical help: our own incredible word wizard, Stephanie Gottwald. The unique illustrations of the brain are the work of former student, Catherine Stoodley, a gifted neuroscientist as well as talented artist at Oxford University. Above all, I want to thank the Center's program coordinator, Wendy Galante, whose continuous, unflagging help and support on this manuscript were inestimable. I could never have written this book without her.

Next, I wish to thank two groups of colleagues with whom I have worked over fifteen years in various capacities: Patricia Bowers, whose work on the double-deficit hypothesis and whose many insights and kindnesses added immeasurably to my scholarship; and my two wonderful colleagues Maureen Lovett and Robin Morris, with whom I worked for ten years on intervention for children with dyslexia. I could not be more grateful for the sustained level of intellectual and personal camaraderie that we and all the members of our three centers shared on these projects. It changed my entire academic direction.

With great continuing gratitude, I wish to thank those foundations and government agencies that have funded various aspects of my research over the last years and made possible many of the insights in this book: National Institute for Child Health and Human Development; Institute for Education Sciences; Haan Foundation for Children; The Dyslexia Research Foundation; Virginia Piper Foundation; Recording for the Blind and Dyslexic; Alden Trust Fund; Stratford Foundation; Tufts Faculty Research grants; and Tisch College for Citizenship and Public Service. I also wish to thank members of my department, the Eliot-Pearson Department of Child Development, and the past and present administrators at Tufts University, particularly Lawrence Bacow, Sol Gittleman, John DiBiaggio, Robert Sternberg, Rob Hollister, and most particularly Wayne Bouchard, for the extraordinary and generous support they provided my Center's research. The collective support of my university and these government agencies and foundations enabled my group to transform a research center into a place where parents, families, and schools in our communities can interact and propel and perform our work on a daily basis. Within this context, I want to thank Anne and Paul Marcus and their foundation for the wonderful and creative ways it has helped our Center thrive and become a place where children, parents, and scholars are equally welcome.

No one person has been more supportive of and generous to the mission of our Center and its research than Barbara Evans, a former teacher and a graduate of the Eliot-Pearson Department of Child Development. She and her husband, Brad Evans, have helped ensure that this direction of research and its application in the community will continue both through our academic scholarship and through the present and future work of Evans Literacy Fellows, a new group of graduate students in the department who pursue reading and language research. The book could never have been completed this year without the personal support and kindness of Barbara Evans and her family.

The same is equally true for Anne Edelstein, my literary agent,

whose larger vision of this project went beyond any work relationship. No one could have been more supportive over seven
years—professionally, conceptually, and personally. Her unflagging faith in me and in the book's potential contribution buoyed
and sustained my spirits through each year and manuscript draft.
Similarly, I thank my editor, Peter Guzzardi, whose wisdom and
knowledge about books and about what is most important in life
helped me to edit this book in such a way as to lose neither the
science nor the spirit of my work. If Peter was the modern-day
Virgil of the project, Gail Winston at HarperCollins was the
book's intellectual Beatrice from the sentence to the overarching
structure. She saw what was there and would not be content until
everyone else also could. I am also most grateful to the superb
and erudite copyediting by Susan Gamer and to David Koral, and
the entire production department at HarperCollins.

Each chapter was read by a different set of researchers, and I
want to acknowledge each of them, but hold none of them responsible. My thanks to:

- The linguist Ray Jackendoff, for his careful help on the first
  chapter (and on a linguistics chapter that disappeared from
  this book but will appear in another).
- David and Amy Abrams and their sons, Daniel and Michael,
  for listening to my first reading of Chapter 1 and getting rid
  of Phineas Gage!
- Barbara Evans, for several insights found by no one else in the
  first chapter.
- My colleague at Tufts the classics scholar Steve Hirsch, for
  providing me with almost a semester-long tutorial on the
  early history section, and for his perspicacious edits in
  Chapters 2 and 3.
- Yori Cohen, the Assyriologist scholar at Tel Aviv University,
  for his generous editing and his help in understanding the
  work on Sumerian pedagogy.
- The Swiss scholar Hans Dahn, for his help with German
  scholarship on the alphabets.
- Pat Bowers and Tami Katzir, for helpful comments on the

whole book, especially the chapters on dyslexia and the developmental chapters.

- The neurologists Al Galaburda and Susana Campasano, for their important edits on the neurological and genetic research sections of Chapters 1 and 8; and geneticist Elena Grigorenko for her comments on Chapter 8.

Shining through the pages of this book is the continuing influence of all my teachers and my family: from my tiny grade school's dedicated nuns, Sisters Salesia, John Vincent, and Rose Margaret, and beloved Sister Ignatius; to Doris Camp in high school; to Father John Dunne, Elizabeth Noel, and Sister Franzita Kane at St. Mary's and Notre Dame; and to Carol Chomsky, Helen Popp, Courtney Cazden, Jeanne Chall, Norman Geschwind, and Martha Denckla at Harvard University. I am especially indebted to Jeanne Chall—the late director of the Harvard Reading Laboratory, my mentor and the person who first gave me Proust's beautiful book *On Reading*, which gave me the germinating thoughts for *Proust and the Squid*.

At least one of these former teachers will appear in some visible or invisible way in every chapter. They are the best teachers one could have in a lifetime. And the best of all my teachers were and are my parents, Frank and Mary Wolf of Eldorado, Illinois. Their way of living every day with unwavering virtue and unspoken generosity gave me and my beloved sister and brothers, Karen, Joe, and Greg, a foundation for life. I will never stop thanking them all.

Finally, I wish to thank my friends, my children, and my husband. This last year and a half was a great test in my life at many levels, and it was only because of my friends and family that ultimately I could make it back to full health and return to this book. In particular, four friends—my sister, Karen Wolf-Smith; Heidi Bally; Cinthia Coletti Haan; and Lynne Tomer Miller—became the human equivalent of angels as they protected, comforted, and guided me back to well-being. I thank them and each of my friends, named and unnamed, from my heart. For my husband, Gil, and my children, Ben and David, I reserve my deepest grati-

tude. Ben and David daily inspired my writing—with fresh anecdotes, and with the reasons why this book must be written by me,
their researcher-mother. They are in every way my life's best light.
And Gil, whose name means "joy" in Hebrew, may the end of this
book bring as much joy (and relief) to you as you have given to me
while writing it! Thank you for every day.

NOTES

Like reading itself, this book is based on hundreds of invisible sources. Because it is a trade book meant for general audiences, I have not given all these sources immediately in a reference or a footnote, as is my way in academic writing. Rather, I have used this notes section with catchphrases rather than numbers for giving the background information. My hope is that the motivated reader who wants to know the source and to dig deeper into the concepts and arguments will use this section along with the text. All the reference materials are found here.

This section provides the chapter, page, and phrase from the text and the sources for references that document it. In some cases, particularly when there are multiple perspectives, I provide further readings.

PREFACE

xi "the acme of a psychologist's achievements"   E. Huey (1968). *The Psychology and Pedagogy of Reading.* Cambridge, Mass.: MIT Press, p. 6.

xiii "I wrote almost all of it in the deepest hope"   M. Robinson (2004). *Gilead.* New York: Farrar, Straus, and Giroux, p. 19.

1 "Words and Music"   J. S. Dunne (2006). A *Vision Quest.* Notre Dame, Ind.: Notre Dame Press, p. viii.

1 "Knowing how something originated"   T. Deacon (1997). *The Symbolic Species.* New York: Norton, p. 23.

CHAPTER 1: READING LESSONS FROM PROUST AND THE SQUID

3 "I believe that reading, in its original essence" M. Proust (1906). *On Reading*, ed. J. Autret and W. Burford (trans. 1971). New York: Macmillan, p. 31.

3 "Learning involves the nurturing of nature" J. LeDoux (2002). *Synaptic Self*. New York: Viking Penguin. p. 9.

3 This plasticity in the brain's design H. J. Neville and D. Bavelier (2000). "Specificity and Plasticity in Neurocognitive Development in Humans." In M. Gazzaniga, ed., *The New Cognitive Neurosciences*. Cambridge, Mass.: MIT Press.

5 When Chinese readers first try to read in English L. H. Tan, J. Spinks, J. Feng, W. Siok, C. Perfetti, J. Xiong, P. Fox, and J. Gao (2003). "Neural Systems of Second Language Reading Are Shaped by Native Language." *Human Brain Mapping*, 18, pp. 158–166.

5 "A biography of any literary person" J. Epstein (1985). "The Noblest Distraction." In J. Epstein, *Plausible Prejudices: Essays on American Writing*. London: Norton.

6 Proust saw reading as a kind of intellectual "sanctuary" Proust, *On Reading*.

6 Scientists in the 1950s Contemporary neuroscientists go well beyond the squid. Today scientists use the sea slug *Aplysia*, the fruit fly, the tiny worm *C. elegans*, and other helpful creatures to learn how neural cells and molecules and genes adjust themselves to the process of learning. Adaptations of these learning processes are implemented in our brains for the purpose of reading.

6 the long central axon of the shy but cunning squid A. L. Hodgkin and A. F. Huxley (1952). "A Quantitative Description of Membrane Current and Its Application to Conduction and Excitation in Nerve." *Journal of Physiology*, 117, pp. 500–544.

6 "There are perhaps no days of our childhood" Proust, *On Reading*, p. 3.

7 This was his sign of respect for the author's gift Letter dated December 10, 1513, from Machiavelli to Francesco Vettori. J. Atkinson and D. Sices, eds. (1996). *Machiavelli and His Friends: Their Personal Correspondence*. Dekalb: Northern Illinois University Press.

7 Machiavelli's tacit understanding   Proust, *On Reading*.

7 the theologian John Dunne   J. Dunne (1973). *Time and Myth*. New York: Doubleday. J. Dunne (1993). *Love's Mind: An Essay on Contemplative Life*. Notre Dame, Ind.: University of Notre Dame Press.

9 the cognitive scientist David Swinney   D. A. Swinney (1979). "Lexical Access during Sentence Comprehension: (Re)considerations of Context Effects." *Journal of Verbal Learning and Verbal Behavior*, 18, pp. 645–659.

9 your semantic and grammatical systems had to function closely with your working memory   A. Baddeley (1986). *Working Memory*. Oxford: Oxford University Press.

11 the first humans who invented writing   S. Dehaene. In K. Fischer and T. Katzir eds. (in press). *Creating Usable Knowledge in Mind, Brain, and Education*. Cambridge: Cambridge University Press.

12 one of the regions of the brain we humans now use for mathematical operations   S. Dehaene (1997). *The Number Sense*. New York: Oxford University Press. S. Dehaene, J. R. Duhamd, M. Aarber, and G. Rozzolatti (2003). *From Monkey Brain to Human Brain*. Cambridge, Mass.: MIT Press.

12 the species' evolutionarily older circuitry   S. Dehaene, H. G. LeClec, J. Poline, D. LeBihan, and L. Cohen (2002). "The Visual Word Form Area: A Prelexical Representation of Visual Words in the Fusiform Gyrus." *Neuroreport*, 13(3), pp. 321–325.

12 "Specialization within a specialization"   T. A. Polk and M. J. Farah (1997). "A Simple Common Contexts Explanation for the Development of Abstract Letter Identities." *Neural Computation*, 9(6), pp. 1277–1289.

12 each neuron in the eye's retina   C. Shatz (2003). "Emergence of Order in Visual System Development." In M. Johnson and Y. Munakata, eds., *Brain Development and Cognition: A Reader*, 2nd ed. Malden, Mass.: Blackwell. C. J. Shatz (1992). "The Developing Brain." *Scientific American*, 267(3), pp. 60–67.

14 the capacity of the neuronal circuits to become virtually automatic D. Hebb (1949). *The Organization of Behavior*. New York: Wiley.

14 learn to "fire together"   Ibid.

14 they create representations   See discussion of mental representations in S. Pinker (1997). *How the Mind Works*. New York: Norton.

14 Stephen Kosslyn   S. M. Kosslyn, N. M. Alpert, W. L. Thompson,
Maljkovie, V. Weise, C. F. Chabris, S. E. Hamilton, S. L. Rauch, and
F. S. Buonanno (1993). "Visual Mental Imagery Activates Topographically
Organized Visual Cortex: PET Investigations." *Journal of Cognitive
Neuroscience*, 5(3), pp. 263–287.

15 connecting how Proust wrote about a single day in his childhood   J. Rewald
(1973). *The History of Impressionism*. New York: Museum of Modern Art.

15 "tell all the Truth, but tell it slant"   E. Dickinson (1961). *The Complete
Poems of Emily Dickinson*, ed. T. J. Johnson. Boston, Mass.: Little, Brown.

16 multiple modalities of text presentation   Throughout this book I will
be presenting a very particular perspective on reading. There is a growing
literature on multiple types of literacy that incorporates a range of different
perspectives on reading in technological formats. See, for example, the
following. G. Kress (2003). *Literacy in the New Media Age*. New York:
Routledge. C. Lewis and B. Fabos (2005). "Instant Messaging, Literacies,
and Social Identities." *Reading Research Quarterly*, 40, pp. 470–501.
D. J. Leu (2000). "Literacy and Technology: Deictic Consequences for Literacy
Education in Our Information Age." In M. Kamil, P. B. Mosenthal,
P. D. Pearson, and R. Barr, eds. *Handbook of Reading Research*. Mahwah,
N.J.: Erlbaum, Vol. 3, pp. 743–770. D. Reinking, M. McKenna, L. Labbo,
and R. D. Kieffer (1998). *Handbook of Literacy and Technology:
Transformations in a Post-Typographic World*. Mahwah, N.J.: Erlbaum.

16 "From so simple a beginning" See a wonderful discussion of the evolution
of these closing words from Darwin's *On the Origin of Species* (1859) in
Sean Carroll (2005), *Endless Forms Most Beautiful*. New York: Norton,
pp. 281–283.

16 reading allows the species to go "beyond the information given"
J. S. Bruner (1973). *Beyond the Information Given*. New York: Norton.

17 one index of the history of thought   Robert Darnton (1986). "A History of
Reading." *Australian Journal of French Studies*, 23, pp. 5–30.

17 more biological and cognitive than cultural-historical   Outside the scope
of this book, there is an extremely rich and expanding literature on the
cultural aspects of literacy and of the multiple literacies that are evolving
today. See the following, and the references they provide. D. Brandt (2000).
Literacies in American Lives. Cambridge: Cambridge University Press. J. Gee
(1996). *Sociolinguistics and Literacies: Ideology in Discourses*. New York:

Falmer. D. Lemonnier-Shallert, and S. Wade (2005). "The Literacies of the Twentieth Century: Stories of Power and the Power of Stories in a Hypertextual World." *Reading Research Quarterly,* 40, pp. 520–529. C. Selfe and G. Hawisher (2004). *Literate Lives in the Information Age: Narratives of Literacy from the United States.* Mahwah, N.J.: Erlbaum.

17 "We feel quite truly that our wisdom begins"   Proust, *On Reading*, p. 35.

19 "Children are wired for sound"   S. Pinker (1997). "Foreword." In D. McGuinness, *Why Our Children Can't Read—And What We Can Do about It: A Scientific Revolution in Reading.* New York: Simon and Schuster.

20 one of the best predictors of later reading   C. Chomsky (1972). "Stages in Language Development and Reading Exposure." *Harvard Educational Review* 42, pp. 1–33. Whitehurst, G. J. Whitehurst, and C. J. Lonigan (2001). "Emergent Literacy: Development from Prereaders to Readers." In S. B. Neuman and D. K. Dickinson, eds., *Handbook of Early Literacy Research.* New York: Guilford, pp. 11–29.

20 a gap of 32 million words   B. Hart and T. Risley (1995). *Meaningful Differences in the Everyday Experience of Young American Children.* Baltimore, Md.: Brookes.

20 who never imagine fighting with dragons   D. Dickinson, M. Wolf, and S. Stotsky (1993). "Words Move: The Interwoven Development of Oral and Written Language in the School Years." In J. Berko-Gleason, ed., *Language Development,* 3rd ed. Columbus, Ohio: Merrill, pp. 369–420.

22 the multidimensioned culture of the Internet   J. Gee (2003). *What Video Games Have to Teach Us about Learning and Literacy.* New York: Palgrave Macmillan. L. A. Henry (2006). "SEARCHing for an Answer: The Critical Role of New Literacies While Reading on the Internet." *Reading Teacher,* 59(7), pp. 614–627. Lewis and Fabos, "Instant Messaging, Literacies, and Social Identities."

22 "It would be a shame"   E. Tenner (2006). "Searching for Dummies." *New York Times*, March 26, p. 12.

CHAPTER 2: HOW THE BRAIN ADAPTED ITSELF TO READ

24 "And so I ambitiously proceed"   A. Manguel (1996). *A History of Reading.* New York: Penguin, p. 22.

24 "The invention of writing"   O. Tzeng and W. Wang (1983). "The First Two R's." *American Scientist*, 71(3), pp. 238–243.

24 may prove to be still earlier signs of the first human efforts to "read" Don Hammill alerted me to this article: M. Balter (2002). "Oldest Art: From a Modern Human's Brow—or Doodling?" *Science*, 295(5553), pp. 247–249.

26 "how it works"   T. Deacon (2002). *The Symbolic Species*. New York: Norton. p. 23.

26 Herodotus tells us   N. Ostler (2005). *Empires of the Word*. New York: Harper, p. 129. See also the historical discussion in V. Fromkin and R. Rodman (1978). *An Introduction to Language*. New York: Holt, Rinehart, and Winston, pp. 20–21.

27 invented only once, or several times   H. Vanstiphout (1996). "Memory and Literacy in Ancient Western Asia." In J. Sasson, ed., *Civilizations of the Ancient Near East*. New York: Simon and Schuster, Vol. 4.

27 "By the mere fact of looking"   Manguel, *A History of Reading*, pp. 27–28.

27 the world of letters may have begun as an envelope for the world of numbers   D. Schmandt-Besserat. "The Earliest Precursor of Writing." *Scientific American*, 1986, pp. 31–40. (Special Issue: Language, Writing, and the Computer.)

28 a series of brain imaging studies S. E. Petersen, P. Fox, M. Posner, M. Minton, and M. Raichle (1989). "Positronemission Tomographic Studies of the Processing of Single Words." *Journal of Cognitive Neuroscience*, 1, pp. 153–170. M. Posner and M. Raichle (1994). *Images of Mind*. New York: Scientific American Library.

29 responsible both for more demanding sensory processing and for making mental representations   S. Pinker (1997). *How the Mind Works*. New York: Norton. (Pinker gives excellent descriptions of representations.)

30 the *angular gyrus* area   N. Geschwind (1977). Lecture, Harvard Medical School.

31 Joseph-Jules Déjerine   N. Geschwind (1974). *Selected Papers on Language and the Brain*. Dordrecht, Netherlands: D. Reidel.

31 pathways to and from the angular gyrus region   J. Demb, R. Poldrack, and J. Gabrieli (1999). "Functional Neuroimaging of Word Processing in Normal

and Dyslexic Readers." In R. Klein and P. McMullen, eds., *Converging Methods for Understanding Reading and Dyslexia*. Cambridge, Mass.: MIT Press.

31 "Have you noticed how picturesque" V. Hugo (1910). *France et Belgique. Alpes et Pyrénées. Voyages et Excursions.*

31 individual Sumerian inscriptions developed into a cuneiform system P. Michalowski (1996). "Mesopotamia Cuneiform: Origin." In P. Daniels and W. Bright, eds., *The World's Writing Systems*. New York: Oxford University Press, pp. 33–36.

32 whose influence continued through the great Akkadian system One scholar, Pitor Michalowski, argues that Sumerian cuneiform was invented "in one fell swoop. . . . [and] was without precedent." Michalowski, "Mesopotamian Cuneiform: Origin."

32 Raulinson dangled on a rope J. DeFrancis (1989). *Visible Speech: The Diverse Oneness of Writing Systems*. Honolulu: University of Hawaii Press, p. 69.

33 So as not to be frustrated by mortal failings Michalowski, "Mesopotamian Cuneiform: Origin."

33 many of the symbols and letters used in writing and numerical systems around the world S. Dehaene (2004). Presentation at the 400th Anniversary of the Vatican Academy of Science, Vatican City, Italy.

35 findings of Raichle's group M. Posner and M. Raichle (1994).

35 John DeFrancis, a scholar of ancient languages and Chinese J. DeFrancis (1989). *Visible Speech: The Diverse Oneness of Writing Systems*. Honolulu: University of Hawaii Press.

36 which Dehaene hypothesizes is the major seat of "neuronal recycling" in literacy S. Dehaene et al. (2002). "The Visual Word Form Area: A Prelexical Representation of Visual Words in the Fusiform Gyrus." B. McCandless, L. Cohen, and S. Dehaene (2003). "The Visual Word Form Area: Expertise for Reading in the Fusiform Gyrus." *Trends in Cognitive Sciences*, 7, pp. 293–299.

37 that is how Chinese symbols are learned by young readers L-H. Tan, J. Spinks, G. Eden, C. Perfetti, and W. Siok. (2005). "Reading Depends on Writing in Chinese." *PNAS*, 102, pp. 8781–8785.

37 their *e-dubba* or "tablet house" schools.   Y. Cohen (2003). "The Transmission and Reception of Mesopotamian Scholarly Texts at the City of Emar." Harvard University (unpublished dissertation.)

38 In current terms, the Sumerians used the first known metacognitive strategy   See the extensive work by Lynne Meltzer and Bethani Roditi and by Maureen Lovett on contemporary uses of metacognitive strategies in instruction. L. Meltzer, L. Pollica, and M. Barzillai (in press). "Creating Strategic Classrooms: Embedding Strategy Instruction in the Classroom Curriculum to Enhance Executive Processes." In L. Meltzer, ed., *Understanding Executive Functioning.* New York: Guilford. L. J. Meltzer, T. Katzir, L. Miller, R. Roddy, and B. Roditi (2004). "Academic Self-Perceptions, Effort, and Strategy Use in Students with Learning Disabilities: Changes over Time." *Learning Disabilities Research and Practice,* 19(2), pp. 99–108. M. Lovett, S. Borden, T. DeLuca, L. Lacerenza, N. Benson, and D. Brackstone (1994). "Training the Core Deficits of Developmental Dyslexia: Evidence of Transfer of Learning after Phonologically and Strategy-based Reading Training Programs." *Developmental Psychology,* 30 (6), pp. 805–822.

38 Without this profoundly important capacity   Steven Pinker discusses this creative feature of language and thought and another equally important combinatorial feature, recursion: "Because human thoughts are combinatorial (simple parts combine) and recursive (parts can be embedded within parts), breathtaking expanses of knowledge can be explored with a finite inventory of mental tools." S. Pinker (1994). *The Language Instinct.* New York: Morrow, p. 360.

39 monkeys have combined the two calls to make a new call   K. Arnold and K. Zuberbuhler (2006). "Language Evolution: Semantic Combinations in Primate Calls." *Nature,* 441(7091), pp. 303–305.

39 cutting-edge reading curricula   M. Wolf, L. Miller, and K. Donnelly (2000). "RAVE-O: A Comprehensive Fluency-Based Reading Intervention Program." *Journal of Reading Disabilities,* 33, pp. 375–386. (Special Issue: The Double-Deficit Hypothesis.)

39 Emesal, or the "fine tongue"   N. Ostler (2005). *Empires of the Word: A Language History of the World.* New York: HarperCollins.

40 "Come sleep, come sleep"   Ibid., pp. 51–52.

40 adopted Sumerian cuneiform script and its teaching methods   J. Pritchard (1969). *Ancient Near East Texts Relating to the Old Testament.* Princeton, N.J.: Princeton University Press.

40 two scribes working intently side by side   Ostler, *Empires of the Word*.

41 For whom have I labored?   J. Maier and J. Gardner, trans. (1981). *Gilgamesh*. New York: Vintage, Book 11.

42 the Native American leader Sequoya   G. T. M. Altmann and A. Enzinger (1997). *The Ascent of Babel: An Exploration of Language, Mind, and Understanding*. New York: Oxford University Press.

42 To illustrate the morphophonemic principle in English   N. Chomsky and M. Halle (1968). *The Sound Pattern of English*. New York: Harper and Row. C. Chomsky (1972). "Stages in Language Development and Reading Exposure." *Harvard Educational Review*, 42, pp. 1–33.

42 English represents a "trade-off"   K. Rayner, B. R. Foorman, C. A. Perfetti, D. Pesetsky, and M. S. Seidenberg (2001). "How Psychological Science Informs the Teaching of Reading." *Psychological Science in the Public Interest*, 2, pp. 31–74.

43 still controversial evidence from German Egyptologists in Abydos   K. T. Zauzich (2001). "Wir alle schreiben Hieroglyphen: Neue Überlegungen zur Herkunft des Alphabets." *Antike Welt*, pp. 167–170.

43 both were considered gifts from the gods   The sacredness of characters is a long tradition found not only in Egypt, Sumer, and China, but also in the kabbalistic traditions of Judaism and in the writing of Islamic prayers.

45 As the linguist Peter Daniels described it   P. Daniels and W. Bright, eds. (1996). *The World's Writing Systems*. New York: Oxford University Press

46 From some 700 standard signs in the Middle Egyptian period   Y. Cohen (2003). Personal correspondence.

47 Cumulative evidence around the world   H. Vanstiphout (1996). "Memory and Literacy in Ancient Western Asia." In J. Sasson, ed., *Civilizations of the Ancient Near East*. New York: Simon and Schuster Macmillan, Vol. 4.

47 This script remains undeciphered   A. Parpola (1994). *Deciphering the Indus Script*. New York: Cambridge University Press.

47 the famous script Linear B   E. Bennett (1996). "Aegean Scripts." In P. Daniels and W. Bright, eds., *The World's Writing Systems*. New York: Oxford University Press, pp. 125–133.

47 a relatively isolated scholar in Stalinist Russia   M. Coe (1992). *Breaking the Mayan Code.* New York: Thames and Hudson.

47 the ancient dyed fibers and twine   J. Quilter and G. Urton (2002). *Narrative Threads: Accounting and Recounting in Andean Khipu.* Austin: University of Texas Press.

48 "The birthing was not a happy event."   DeFrancis, *Visible Speech,* p. 93.

49 Gish Jen   Personal correspondence (March 2004).

49 Lisa See's novel   L. See (2005). *Snow Flower and the Secret Fan.* New York: Random House.

50 For centuries this remarkable writing system   C. Simon (2005). "Novel's Powerful Prose Brings History to Life." *Boston Globe,* July 27.

CHAPTER 3: THE BIRTH OF AN ALPHABET AND SOCRATES' PROTESTS

51 ringed by the wine-dark sea   Homer, *Odyssey,* Book 19. R. Eagles (trans. 1990). New York: Renguin, lines 194–199.

51 the Egyptologists John Darnell and Deborah Darnell   J. Darnell and D. Darnell (2002). *Theban Desert Road Survey in the Egyptian Western Desert.* Chicago, Ill.: Oriental Institute of the University of Chicago. J. N. Wilford (1999). "Finds in Egypt Date Alphabet in Earlier Era." *New York Times,* November 14, Section 1, p. 1.

52 "clearly the oldest of alphabetic writing"   Wilford, "Finds in Egypt."

52 bustling with trade from both ships and overland caravans   W. Whitt (1996). "The Story of the Semitic Alphabet." In J. Sasson, ed., *Civilizations of the Ancient Near East.* New York: Simon and Schuster, Vol. 4.

52 In Ugarit, different peoples spoke at least ten languages   A. Robinson (1995). *The story of Writing.* London: Thames and Hudson.

52 independent consonant signs were combined with consonant signs that distinguished adjacent vowels   P. Daniels and W. Bright, eds. (1996). *The World's Writing Systems.* New York: Oxford University Press.

53 Ugaritic writing would be considered an *abjad*   By contrast, some classicists, such as Eric Havelock, consider it a syllabary. The fact that Ugaritic

is classified in two different ways reflects its linkage with both types of systems, much like the earlier Wadi el-Hol script.

53 the oral and written Ugaritic language influenced the writing of the Hebrew Bible.   Yori Cohen (2000). Personal correspondence, January 9.

53 A few scholars, including the biblical scholar James Kugel   J. Kugel (2003). *The God of Old: Inside the Lost World of the Bible*. New York: Free Press.

53 There is a relevant biblically inspired short story   T. Mann, "Das Gesetz." *Collected Stories of Thomas Mann* (1943/1966). Katia Mann, ed. (*Sämmtliche Erzählungen, Band I*). Frankfurt, Germany: S. Fischer Verlag, pp. 329–395.

55 a complete correspondence between each phoneme in the language and each visual sign or letter   E. Havelock (1976). *Origins of Western Literacy*. Ontario, Canada: Ontario Institute for Studies in Education.

55 the Greek system (750 BCE) was the first to satisfy all conditions   I. Gelb (1963). *A Study of Writing*, 2nd ed. Chicago, Ill.: University of Chicago Press.

55 The Assyriologist Yori Cohen   Y. Cohen (2000). Personal correspondence, January 9.

57 Ventris never expected to decipher colloquial Greek   Steve Hirsh (2004). Personal correspondence. See also J. Chadwick (1958). *The Decipherment of Linear B*. Cambridge: Cambridge University Press

57 an educated Greek citizen committed to memory   W. Ong (1982). *Orality and Literacy*. London: Methuen. R. Scott (2003). *The Gothic Enterprise*. Berkeley: University of California Press.

57 The scholar Millman Perry Cited in Ong, *Orality and Literacy*.

57 these formulas enabled ancient Greeks to memorize and recite   Scott, *The Gothic Enterprise*.

58 They remind us of the significant effects of culture   L. Hirschfeld and S. Gelman (1994). *Mapping the Mind: Domain Specificity in Cognition and Culture*. New York: Cambridge University Press.

58 Several scholars suggest that the Greek written alphabet came into existence   B. Powell (1991). *Homer and the Origin of the Greek Alphabet*. Cambridge: Cambridge University Press.

59 The Phoenicians, in turn, based their letters   G. Sampson (1985). *Writing Systems.* London: Hutchinson.

59 *alpha* and *beta* come from the Phoenician *aleph* and *bet*   Ibid.

59 At least one quiet war is being waged   K. T. Zauzich (2001). "Wir alle schreiben Hieroglyphen: Neue Uberlegungen zur Herkunft des Alphabets." *Antike Welt,* pp. 167–170.

59 what the German scholar Joseph Tropper calls the "standard theory" J. Tropper (2001). "Entstehung und Fruhgeschichte des Alphabets." *Antike Welt,* 32(44), pp. 353–358.

59 These two writing systems must have had a shared Semitic mother Zauzich, "Wir alle schreiben Hieroglyphen," p. 167.

59 at least one version ends with Cadmus strewing bloodied teeth   R. Graves (1955). *Greek Myths.* New York: George Braziller.

60 Various influential twentieth-century scholars   Havelock, *Origins of Western Literacy.*

60 the alphabet's increased efficiency over other systems   Daniels and Bright, *The World's Writing Systems.*

60 the alphabet's facilitation of novel thoughts   Havelock, *Origins of Western Literacy.*

61 brain images of modern Chinese   D. Bolger, C. Perfetti, and W. Schneider (2005). "Cross-Cultural Effect on the Brain Revisited: Universal Structures Plus Writing System Variation." *Human Brain Mapping,* 25, pp. 92–104.

61 This differential use of hemispheres   R. S. Lyman, S. T. Kwan, and W. H. Chao (1938). "Left Occipito–Parietal Brain Tumor with Observations on Alexia and Agraphia in Chinese and in English." *Chinese Medical Journal,* 54, pp. 491–515.

62 Note that Figure 3-1 shows the brain of a Japanese reader proficient in both Japanese scripts.   The similarities between the syllabary and the alphabet-reading brains appear most apparent in discrete portions of the frontal and temporal lobes. These particular areas subserve phonological processes that range from sound identification in words like "spaghetti" to figuring out the stress pattern in words like "despot." These regions activate more prominently in both the syllabary and the alphabet reader, because both these writing

systems need a great deal of "up-front" time to process the tiny phonemes and larger syllables within words. As seen in Figure 3–1, a very important region in the frontal lobe (called Broca's area) also has specific areas of specialization that contribute to the brain's efficiency: some for phonemes in words and some for meanings. An analogous, multifunction region in the upper temporal lobes and lower, adjacent parietal lobes also appears involved in sound analysis, as well as in the meanings of words. Again, these two regions for sound and meaning analysis show more expansive activation in the alphabet and syllabary-reading brains than for Chinese readers.

63 Japanese readers use pathways similar to those of the Chinese
K. Nakamura et al. (2002). "Modulation of the Visual Word Retrieval System in Writing: A Functional MRI Study on the Japanese Orthographies." *Journal of Cognitive Neuroscience,* 14, pp. 104–115.

63 the same words written in kana   L. B. Feldman and M. T. Turvey (1980). "Words Written in Kana are Named Faster Than the Same Words Written in Kanji." *Language and Speech,* 23, pp. 141–147.

63 cognitive scientists from the University of Pittsburgh   D. Bolger, C. Perfetti, and W. Schneider (2005). "Cross-Cultural Effect on the Brain Revisited: Universal Structures Plus Writing System Variation." *Human Brain Mapping,* 25, pp. 92–104.

63 the "universal reading system"   Ibid.

64 Factors like the number of symbols   As we saw earlier with regard to the Chinese logosyllabary-reading brain, the amount of time pupils in ancient Sumer and modern China expend on drawing each character in order to learn it is visibly reflected in their brain activation patterns. Chinese readers' motoric memory areas in the frontal lobe activate every time they read.

64 Philosophers including Benjamin Whorf   According to Whorf, different languages can have words which influence how we think so powerfully that our concepts of the named thing in that language cannot be understood in another language: e.g.: the various words for specific types of snow used in an Aleutian language. See the excellent discussion in R. Jackendoff (2002). *Foundations of Language.* Oxford: Oxford University Press, pp. 292–293.

64 Walter Benjamin   The early-twentieth-century German philosopher Walter Benjamin speculated lyrically about what this difference might mean: "The power of a country road is different when one is walking along it from when one is flying over it by airplane. In the same way, the power of a text is different when

it is read from when it is copied out. The airplane passenger sees only how the road pushes through the landscape, how it unfolds according to the same laws as the terrain surrounding it. Only he who walks the road on foot learns of the power it commands, and of how, from the very scenery that for the flier is only the unfurled plain, it calls forth distances, belvederes, clearings, prospects at each of its turns like a commander deploying soldiers at a front. Only the copied text thus commands the soul of him who is occupied with it, whereas the mere reader never discovers the new aspects of his inner self that are opened by the text, that road cut through the interior jungle forever closing behind it: because the reader follows the movement of his mind in the free flight of daydreaming, whereas the copier submits it to command. The Chinese practice of copying books was thus an incomparable guarantee of literary culture." W. Benjamin (1978). *Reflections.* New York: Harcourt Brace Jovanovich, p. 66.

64 Guinevere Eden observed   G. Eden (2000). Presentation at Fluency Conference, Dyslexia Research Foundation, Crete.

65 The classicist Eric Havelock   To Havelock, oral culture "had hitherto placed severe limitations upon the verbal arrangement of what might be said, or thought. More than that, the need to remember had used up a degree of brain-power-of psychic energy—which now was no longer needed. . . . The mental energies thus released, by this economy of memory, have probably been extensive, contributing to an immense expansion of knowledge available to the human mind." *Origins of Western Literacy*, p. 49.

65 the psychologist David Olson   D. Olson (1977). "From Utterances to Text: The Bias of Language in Speech and Writing." *Harvard Educational Review*, 47(3), pp. 257–281.

65 Lev Vygotsky said   L. Vygotsky (1962). *Thought and Language.* Cambridge, Mass.: MIT Press.

67 research by the speech scientist Grace Yeni-Komshian   G. Yeni-Komshian and H. Bunnell (1998). "Perceptual Evaluations of Spectral and Temporal Modifications of Deaf Speech." *Journal of the Acoustical Society of America*, 104(2), pp. 637–647. G. Yeni-Komshian (1998). "Speech Perception." In J. B. Gleason and N. Ratner, eds., *Psycholinguistics*. New York: Harcourt.

67 because all sounds are co-articulated   D. Shankweiler and I. Liberman (1972). "Misreading: Searching for Causes." In J. Kavanagh and I. Mattingly, eds., *Language by Ear and by Eye*. Cambridge, Mass.: MIT Press, pp. 293–329.

67 "One of the greatest challenges"   Yeni-Komshian, "Speech Perception."

68 the Greek alpha for the vowel "a" emerged from the Phoenician word *aleph*   P. Swiggers (1996). *Ancient Grammar: Content and Context.* Leuven, Belgium: Peeters.

68 why slightly differing scripts appear in different Greek cities   L. Threatte (1996). "The Greek Alphabet." In P. Daniels and W. Bright, eds., *The World's Writing Systems.* New York: Oxford University Press, pp. 271–280.

68 Changing the letters of a writing system to match a local dialect "This structural innovation was a major step in the history of writing: it made possible the exhaustive representation of the linear sequence of the *sound segments* constituting a message, and thus allowed the direct continuous reading of any text, not requiring any grammatical information to be supplied by the reader."   Swiggers, *Ancient Grammar: Content and Context,* p. 265.

68 educated Greeks considered their highly developed oral culture   Havelock, *Origins of Western Literacy.*

69 "Socrates himself wrote nothing at all"   M. Nussbaum (1997). *Cultivating Humanity: A Classical Defense of Reform in Liberal Education.* Cambridge, Mass.: Harvard University Press, p. 34.

69 "It is not too much to say that with Aristotle"   F. G. Kenyon (1932). *Books and Readers in Ancient Greece and Rome.* Oxford: Clarendon Press, p. 25.

69 the "stinging gadfly"   Plato. "Apology." In E. Hamilton and H. Cairns, eds. (1961). *The Collected Dialogues.* Princeton, N.J.: Princeton University Press, pp. 30E–31A.

70 Aristotle was already immersed in "the habit of reading"   Ong, *Orality and Literacy.* Kenyon, *Books and Readers in Ancient Greece and Rome,* p. 25.

70 This intensive mode of learning   The questioning of traditional beliefs had antecedents in pre-Socratic and Sophist thinkers. These Greek teachers in the second half of the fifth century BCE taught rhetoric and logic to the wealthy citizenry, along with a way of thinking that sought to distinguish universal values from culturally created beliefs. In Aristophanes' comedy *The Clouds,* Socrates is sardonically portrayed as an out-of-control Sophist, a caricature dismissed by Socrates and Plato.

71 "with all our intelligence"   A. Rich (1978). *The Dream of a Common Language.* New York: Norton.

71 "If I tell you that this is the greatest good"   Plato, "Apology," p. 38A.

72 "The way of words"   J. Dunne (1993). *Love's Mind: An Essay on Contemplative Life*. Notre Dame, Ind.: University of Notre Dame Press, p. 31.

72 In the film *The Paper Chase*   *The Paper Chase* (1973). Directed by James Bridges, Twentieth Century Fox.

72 . . . unlike the "dead discourse" of written speech . . .   Plato, "Phaedrus," p. 274.

73 Vygotsky described the intensely generative relationships   Vygotsky, *Thought and Language*.

74 they "seem . . . as though they were intelligent"   Plato, "Phaedrus." In E. Hamilton and H. Cairns, eds. (1961). *The Collected Dialogues*. Princeton, N.J.: Princeton University Press, p. 275d.

74 "[In] modern Guatemala"   N. Ostler (2005). *Empires of the Word: A Language History of the World*. New York: HarperCollins, p. 85.

74 "If men learn this"   Plato, "Phaedrus," pp. 274d, e. This quote comes at the end of a wonderful parable where Socrates set his sights on the preservation of what individual memory ultimately allowed. Note that in this passage, Socrates collapsed various aspects of memory that modern scholars would carefully distinguish from each other: the role of memorization as a tool in education; the preservation of an individual's long-term memory capacities; and the preservation of collective or cultural memory in each person. The complete passage is instructive:

> The story is that in the region of Naucratis in Egypt there dwelt one of the old gods of the country, the god to whom the bird called Ibis is sacred, his own name being Theuth. He it was that invented number and calculation, geometry and astronomy, not to speak of draughts and dice, and above all writing. Now the king of the whole country at that time was Thamus. To him came Theuth, and revealed his arts, saying that they ought to be passed on to the Egyptians in general. Thamus asked what was the use of them all, and condemned what he thought the bad points and praised what he thought the good. On each art, we are told, Thamus had plenty of views both for and against. But when it came to writing, Theuth said, "Here, O king, is a branch of learning that will make the people of Egypt wiser and improve their memories; my discovery provides a recipe for memory and wisdom."
> But the king answered and said, "O man full of arts, to one it is given to create the things of art, and to another to judge what measure of harm and of profit they have for those that shall employ them. And

so it is that you, by reason of your tender regard for the writing that is your offspring, have declared the very opposite of its true effect. If men learn this, it will implant forgetfulness in their souls; they will cease to exercise memory because they rely on that which is written, calling things to remembrance no longer from within themselves, but by means of external marks. What you have discovered is a recipe *not for memory, but for reminder.*"

76 "like papyrus rolls"   Plato, "Protagoras," p. 329a.

76 "Once a thing is put in writing"   Ibid.

77 Underneath his ever-present humor and seasoned irony   I am grateful to Steve Hirsh for helping me see the intended humor in many of the passages I use here.

78 "with all our intelligence"   Rich, *The Dream of a Common Language.*

78 "youngest members of our species"   This is a phrase often used endearingly by the late child linguist Mathilde Holzman of Tufts.

CHAPTER 4: THE BEGINNINGS OF READING DEVELOPMENT, OR NOT

79 "Among the many worlds"   H. Hesse (1931). "The Magic of the Book." In S. Gilbar ed., *Reading in Bed.* Jaffrey, N.H.: Godine. First published in *Bücherei und Bildungspflege.* Stettin, (1931). p. 305.

81 "When the first baby laughed"   J. M. Barrie (1904). *Peter Pan.* New York: Scribner, p. 36.

81 "It seems to me that"   K. Chukovsky and M. Morton (1963). *From Two to Five.* Berkeley: University of California Press, p. 7.

82 Decade after decade of research   C. Chomsky (1972). "Stages in Language Development and Reading Exposure." *Harvard Educational Review,* 42, pp. 1–33. C. Snow, P. Griffen, and M. S. Burns, eds. (2005). *Knowledge to Support the Teaching of Reading: Preparing Teachers for a Changing World.* San Francisco, Calif.: Jossey-Bass. G. J. Whitehurst, and C. J. Lonigan (2001). "Emergent Literacy: Development from Prereaders to Readers." In S. B. Neuman and D. K. Dickinson, eds., *Handbook of Early Literacy Research.* New York: Guilford, pp. 11–29.

82 How a child first learns to read   P. McCardle, J. Cooper, G. Houle, N. Karp, and D. Paul Brown (2001). "Emergent and Early Literacy: Current Status and Research Directions." *Learning Disabilities Research and Practice,* 16(4). (Special issue.) See also the following. E. D. Hirsch (2003). "Reading

Comprehension Requires Knowledge of the Words and the World." *American Educator,* 27(10,12), pp. 1316–1322, 1328–1329, 1348. S. Neuman (2001). "The Role of Knowledge in Early Literacy." *Reading Research Quarterly,* 36, pp. 468–475.

82 the importance of touch   T. Field (2000). *Touch Therapy.* New York: Churchill Livingstone.

82 In a zany, endearing scene   *Three Men and a Baby* (1987). Leonard Nimoy, director, Touchstone Pictures.

82 captured the imagination of millions of children   M. W. Brown (1947). *Goodnight Moon.* New York: Harper and Row.

83 All these reasons and more provide an ideal beginning   D. Dickinson, M. Wolf, and S. Stotsky (1992). "Words Move: The Interwoven Development of Oral and Written Language in the School Years." In J. B. Gleason, ed., *The Development of Language,* 3rd ed. New York: Macmillan. R. New (2001). "Early Literacy and Developmentally Appropriate Practice: Rethinking the Paradigm." In S. B. Neuman and D. P. Dickinson, *Handbook of Early Literacy.* New York: Guilford, pp. 245–263. P. McCardle and V. Chhabra (2004). *The Voice of Evidence in Reading Research.* Baltimore, Md.: Brookes.

83 What the ancient writers of the Rig Veda recognized   N. Ostler (2005). *Empires of the Word: A Language History of the World.* New York: HarperCollins.

84 a major cognitive change also has begun connecting   D. M. Pease, J. B. Gleason, and B. A. Pan (1993). "Learning the Meaning of Words: Semantic Development and Beyond." "In J. B. Gleason (ed.), *The Development of Language,* 3rd ed. New York: Macmillan.

84 The more children are read to   J. Frijters, R. Barron, and M. Brunello (2000). "Child Interest and Home Literacy as Sources of Literacy Experience: Direct and Mediated Influences on Letter Name and Sounds Knowledge and Oral Vocabulary." *Journal of Educational Psychology,* 92(3), pp. 466–477. G. J. Whitehurst and C. J. Lonigan (1998). "Child Development and Emergent Literacy." *Child Development,* 69(3), pp. 848–872.

84 The cognitive scientist Susan Carey   S. Carey (2004). "Bootstrapping and the Origin of Concepts." *Daedalus,* 133, pp. 59–68.

84 the child's "linguistic genius"   K. Chukovsky and M. Morton (1963). *From Two to Five.* Berkeley: University of California Press.

84 Phonological development   S. Brady (1991). "The Role of Working
Memory in Reading Disability." In S. Brady and D. Shankweiler, eds.,
Phonological Processes in Literacy: A Tribute to Isabelle Liberman. Hillsdale,
N.J.: Lawrence Erlbaum, pp. 129–152.

84 Semantic development   J. Anglin (1993). "Vocabulary Development:
A morphological analysis." *Monographs of the Society for Research in Child
Development,* 58(10), pp. 1–166.

84 Syntactic development   A. Charity, H. Scarborough, and P. Griffin (2003).
"Familiarity with School English in African-American Children and Its
Relation to Reading Achievement." *Child Development,* 75, pp. 1340–1356.

85 Morphological development   J. Berko (1958). "The Child's Learning of
English Morphology." *Word,* 14, pp. 150–177. R. Brown (1973). *A First
Language: The Early Stages.* Cambridge, Mass.: Harvard University Press.
J. G. Devilliers and P. A. Devilliers (1973). "A Cross-Sectional Study of the
Acquisition of Grammatical Morphemes in Child Speech." *Journal of
Psycholinguistic Research,* 2, pp. 267–278.

85 pragmatic development   C. Gidney (2002). "The Child as Communicator."
In Tufts Faculty of the Eliot Pearson Department of Child Development, *Pro-
Active Parenting.* New York: Berkley, pp. 241–265. A. S. Ninio and C. E. Snow
(1996). *Pragmatic Development.* Boulder, Colo.: Westview.

86 Jean Piaget, described children as egocentric at this age   J. Piaget (1926).
The Language and Thought of the Child. London: Routledge and Kegan Paul.

86 It is their gradually evolving ability to think   See the following works on
children's theory of mind. P. C. Fletcher, F. Happe, U. Frith, S. C. Baker,
R. J. Dolan, R. S. Frackowiak, and C. D. Frith (1995, Nov.) "Other Minds
in the Brain: A Functional Imaging Study of 'Theory of Mind' in Story
Comprehension." *Cognition,* 57(2), pp. 109–128. M. D. Hauser, and
E. Spelke (2004). "Evolutionary and Developmental Foundations of Human
Knowledge." In M. Gazzaniga, ed., *The Cognitive Neurosciences.* Cambridge,
Mass.: MIT Press, Vol. 3. S. Baron-Cohen, H. Tager-Flusberg, and D. Cohen,
eds. (2000). *Understanding Other Minds,* 2nd ed. Oxford: Oxford University
Press.

86 series of books by Arnold Lobel.   A. Lobel (1970). *Frog and Toad Are
Friends.* New York: HarperCollins.

86 In James Marshall's famous series   J. Marshall (1972). *George and
Martha.* New York: Houghton Mifflin.

87 First, and most obviously, the special vocabulary   C. Pappas and E. Brown (1987). "Learning to Read by Reading: Learning How to Extend the Functional Potential of Language." *Research on the Teaching of English*, 21(2), pp. 160–177. V. Purcell-Gates, E. McIntyre, and P. Freppon (1995). "Learning Written Storybook Language in School: A Comparison of Low-SES Children in Skills-Based and Whole-Language Classrooms." *American Educational Research Journal*, 32(3), pp. 659–685.

87 Indeed, by kindergarten   A. Biemiller (1977). "Relationship between Oral Reading Rate for Letters, Words, and Simple Text in the Development of Reading Achievement." *Reading Research Quarterly*, 13, pp. 223–253. A. Biemiller (1999). *Language and Reading Success*. Cambridge, Mass.: Brookline. J. B. Gleason, ed. (1993). *The Development of Language*, 3rd ed. New York: Macmillan.

87 A large portion of these thousands of words   Anglin, "Vocabulary Development: A Morphological Analysis."

88 Few children under age five hear "for" used   C. Peterson and A. McCabe (1991). "On the Threshold of the Story Realm: Semantic versus Pragmatic Use of Connectives in Narratives." *Merrill-Palmer Quarterly*, 37(3), pp. 445–464.

88 Studies by reading researcher Victoria Purcell-Gates   V. Purcell-Gates (1986). "Three Levels of Understanding about Written Language Acquired by Young Children Prior to Formal Instruction." In J. Niles and R. Lalik, eds., *Solving Problems in Literacy*. Rochester, N.Y.: National Reading Conference. V. Purcell-Gates (1988). "Lexical and Syntactic Knowledge of Written Narrative Held by Well-Read-To Kindergartners and Second-Graders." *Research in the Teaching of English*, 22(2), pp. 128–160.

88 One recent study by the sociolinguist Anne Charity   A. Charity, H. Scarborough, and Griffin, P. (2003). "Familiarity with School English in African-American Children and Its Relation to Reading Achievement." H. Scarborough, W. Dobrich, and M. Hager (1991). "Preschool Literacy Experiences and Later Reading Achievement." *Journal of Learning Disabilities*, 24(8), pp. 508–511.

89 an extremely important, largely invisible aspect of intellectual development   D. Gentner and M. Rattermann (1991). "Language and the Career of Similarity." In A. Gelman and J. P. Byrnes, eds., *Perspectives on Language and Thought: Interrelations in Development*. Cambridge: Cambridge University Press, pp. 225–277.

89 A charming example of early analogical skills   H. A. Rey (1941). *Curious George*. New York: Houghton Mifflin.

89 This kind of cognitive information is part of what goes into "schemata" W. Kintsch and E. Greene (1978). "The Role of Culture-Specific Schemata in the Comprehension and Recall of Stories." *Discourse Processes,* 1(1), pp. 1–13.

90 According to some researchers, being read to is only part   Biemiller, *Language and Reading Success.* Scarborough et al., "Preschool Literacy Experiences and Later Reading Achievement." J. Frijters, R. Barron, and M. Brunello (2000). "Child Interest and Home Literacy as Sources of Literacy Experience: Direct and Mediated Influences on Letter Name and Sounds Knowledge and Oral Vocabulary." *Journal of Educational Psychology,* 92(3), pp. 466–477.

91 The answer is in the Notes section for this page. Yes, they are the same characters.

92 Susan Carey's notion of "bootstrapping"   S. Carey (2004). "Bootstrapping."

92 At a simple level   Bookheimer and his colleagues show that naming objects involves a subset of reading processes. This is exactly right, but the differences between naming objects and letter naming letters show more. See my discussion on this topic in Chapter 7. S. Y. Bookheimer, T. A. Zeffiro, T. Blaxton, W. Gaillard, W. Theodore (2004). "Regional Cerebral Blood Flow during Object Naming and Word Reading." *Human Brain Mapping,* 3(2), pp. 93–106.

93 the quintessential human activity   W. Benjamin (1978). *Reflections,* trans. Edmund Jepheott, ed. P. Demetz. New York: Harcourt and Brace.

93 familiar words and signs in the child's environment   D. Dickinson, M. Wolf, and S. Stotsky (1993). "Words Move: The Interwoven Development of Oral and Written Language in the School Years." In J. Berko-Gleason, ed., *Language Development,* 3rd ed. Columbus, Ohio: Merrill, pp. 369–420.

94 a "logographic" stage   L. C. Ehri (1997). "Sight Word Learning in Normal Readers and Dyslexic." In B. A. Blachman, ed., *Foundations of Reading Acquisition and Dyslexia: Implications for Early Intervention.* Mahwah, N.J.: Lawrence Erlbaum, pp. 163–189.

94 Twenty-six years ago, a colleague of mine at Tufts   D. Elkind (1981). *The Hurried Child.* Boston, Mass.: Addison-Wesley.

94 The growth of myelin follows a developmental schedule   P. Yakovlev and A. Lecours (1967). "The Myelogenetic Cycles of Regional Maturation of the Brain." In A. Minkowski, ed., *Regional Development of the Brain in Early*

Life. Oxford: Blackwell Scientific. C. A. Nelson and M. Luciana, eds. (2001). *Handbook of Developmental Cognitive Neuroscience*. Cambridge, Mass.: MIT Press.

95 The behavioral neurologist Norman Geschwind   N. Geschwind (1965). "Disconnexion Syndrome in Animals and Man (Parts 1 and 2)." *Brain*, 88, pp. 237–294.

95 To be sure, our own research on language   M. Wolf and D. Gow (1985). "A Longitudinal Investigation of Gender Differences in Language and Reading Development." *First Language*, 6, pp. 81–110.

95 The British reading researcher Usha Goswami   U. Goswami (2004). Comments at Mind, Brain, and Education Conference. Harvard University.

96 potentially counterproductive for many children   Elkind, *The Hurried Child*.

96 "As I read the alphabet"   H. Lee, (1960). *To Kill a Mockingbird*. New York: Warner, pp. 17–18.

96 "They obliged me completely and all at once."   P. Fitzgerald (2004). "Schooldays." In T. Dooley, ed., *Afterlife*. New York: Counterpoint. Quoted in Katharine Powers (2003). "A Reading Life." *Boston Globe*, November 16, p. H9.

97 Glenda Bissex provides   G. L. Bissex (1980). *Gnys at Work: A Child Learns to Write and Read*. Cambridge, Mass.: Harvard University Press.

97 the complex notion that letters correspond to the sounds inside words.   Dickinson et al., "Words Move: The Interwoven Development of Oral and Written Language in the School Years," pp. 369–420.

98 called "invented spelling" by Carol Chomsky and Charles Read   Chomsky, "Stages in Language Development and Reading Exposure." C. Read (1971). "Preschool Children's Knowledge of English Phonology." *Harvard Educational Review*, 41, pp. 1–54.

98 This spelling has been found in children's writing   Dickinson et al., "Words Move."

98 a wonderful complement to the actual reading process.   M. Pressley (1998). *Reading Instruction That Works: The Case for Balanced Teaching*. New York: Guilford.

99 the reading expert Marilyn Adams   M. Adams (1990). *Beginning to Read.* Cambridge, Mass.: MIT Press.

99 one of the two best predictors of later reading achievement A. Burhanpurkar and R. Barron (1997). "Origins of Phonological Awareness Skill in Pre-Readers: Roles of Language, Memory, and Proto-Literacy." Paper presented at Society for Research in Child Development meeting, Washington, D.C., April. A. Bus and M. Ijzendoorn (1999). "Phonological Awareness and Early Reading: A Meta-Analysis of Experimental Training Studies." *Journal of Educational Psychology*, 91(3), pp. 403–414. L. C. Moats (2000). *Speech to Print: Language Essentials for Teachers.* Baltimore, Md.: Brookes. Scarborough et al., "Preschool Literacy Experiences and Later Reading Achievement."

100 Lynne Bradley and Peter Bryant   L. Bradley and P. E. Bryant (1983). "Categorizing Sounds and Learning to Read—A Causal Connection." *Nature*, 301, pp. 419–421. L. Bradley and P. E. Bryant (1985). *Rhyme and Reason in Spelling.* Ann Arbor: University of Michigan Press. P. E. Bryant, M. MacLean, and L. Bradley (1990). "Rhyme, Language, and Children's Reading." *Applied Psycholinguistics*, 11(3), pp. 237–252.

100 wordplay, jokes, and songs   Brady, "The Role of Working Memory in Reading Disability." R. Stacey (2003). *Thinking About Language: Helping Students Say What They Mean and Mean What They Say.* Cambridge, Mass.: Landmark School.

100 The Scottish language researcher Katie Overy   K. Overy (2003). "Dyslexia and Music: From Timing Deficits to Musical Intervention." *Annals of the National Academy of Science*, 999, pp. 497–505. K. Overy, A. C. Norton (2004). K. T. Cronm, N. Gaab, D. C. Alsop, E. Winner, and G. Schlaug (August 2004). "Imaging Melody and Rhythm Processing in Young Children: Auditory and Vestibular Systems." *NeuroReport*, 15(11), pp. 1723–1726.

100 Catherine Moritz   C. Moritz (2007). "Relationships between Phonological Awareness and Musical Rhythm Subskills in Kindergarten Children." Master's thesis, Tufts Universtiy.

101 These seemingly simple methods   B. Blachman, E. Ball, R. Black, and D. Tangel (2000). *Road to the Code.* Baltimore, Md.: Brookes. B. Foorman, D. Francis, D. Winikates, P. Mehta, C. Schatschneider, and J. Fletcher (1997). "Early Intervention for Children with Reading Disabilities." *Scientific Studies of Reading*, 1(3), pp. 255–276.

101 The reading researcher Louisa Cook Moats   Moats, *Speech to Print: Language Essentials for Teachers.* L. Moats (2003). *LETRS. Language*

Essentials for Teachers of Reading and Spelling, Preliminary Version, Book 3, Modules 7, 8, 9: Foundations for Reading Instruction. Longmont, Cdo.: Sopris West Educational Services.

101 Systematic tools   The researchers Marilyn Jager Adams, Susan Brady, Benita Blachman, and Louisa Cook Moats provide practical guidelines and sensible cautions about how critical research should be applied. Adams, in her comprehensive book *Beginning to Read*, helped usher in an important trend— giving phoneme awareness tests to all kindergartners. She cautioned, however, that some educators mistakenly hold children back from grade 1 if they fail these tasks. Because phoneme awareness skills take a while to develop and are aided by learning to read, holding a child back on the basis of phoneme awareness abilities makes no sense at all. See also other work by Benita Blachman, Ed Kame'enui, and Deborah Simmons, and a program review in Sally Shaywitz's *Overcoming Dyslexia*. Another great resource is Robbie Stacy's book on language games, *Thinking About Language*.

102 a chilling finding   B. Hart and T. Risley (2003). "The Early Catastrophe." *American Educator*, 27(4), pp. 6–9. T. Risley and B. Hart (1995). *Meaningful Differences in the Everyday Experiences of Young American Children*. Baltimore, Md.: Brookes.

102 What Louisa Cook Moats calls "word poverty"   L. C. Moats (2001). "Overcoming the Language Gap." *American Educator*, 25(5), pp. 8–9.

103 In a survey of three communities in Los Angeles   C. Smith, R. Constantino, and S. Krashen (1997). "Differences in Print Environment for Children in Beverly Hills, Compton, and Watts." *Emergency Librarian*, 24(4), pp. 8–9.

103 The psychologist Andrew Biemiller   Biemiller, *Language and Reading Success*.

103 The interrelatedness of vocabulary development and later reading comprehension   K. Stanovich (1986). "Matthew Effects in Reading: Some Consequences of Individual Differences in the Acquisition of Literacy." *Reading Research Quarterly*, 21(4), pp. 360–407. A. Cunningham and K. Stanovich (1993). "Children's Literacy Environments and Early Word Recognition Subskills." *Reading and Writing: An Interdisciplinary Journal*, 5, pp. 193–204. A. Cunningham and K. Stanovich (1998). "What Reading Does for the Mind." *American Educator*, 22, pp. 8–15.

103 the educator Catherine Snow of Harvard   C. Snow (1996.) Quoted in Kate Zernike (1996). "Declining Art of Table Talk a Key to Child's Literacy." *Boston Globe*, pp. 1, 30, January 15.

104 As the policy maker Peggy McCardle notes   P. McCardle and V. Chhabra, eds. (2004). *The Voice of Evidence in Reading Research*. Baltimore, Md.: Brookes.

106 Examining the many issues swirling around bilingualism   D. August and K. Hakuta (1997). *Improving Schooling for Language- Minority Children*. Washington, D.C.: National Academies Press. Center for Applied Linguistics (2003). "Development of English Literacy in Spanish-Speaking Children: A Biliteracy Research Initiative Sponsored by the National Institute of Child Health and Human Development and the Institute of Education Sciences of the Department of Education." From www.cal.org/delss. B. R. Foorman, C. Goldenberg, C. D. Carlson, W. Saunders, and S. D. Pollard-Durodola (2004). "How Teachers Allocate Time During Literacy Instruction in Primary-Grade English Language Learner Classrooms." In P. McCardle and V. Chhabra, eds., *The Voice of Evidence in Reading Research*. Baltimore, Md.: Brookes, pp. 289–328.

106 English-language learners who know a concept or word   J. Chall (1983). *Stages of Reading Development*. New York: McGraw-Hill. August et al., *Improving Schooling for Language-Minority Children*.

106 Little is more important to learning to read English   M. Collins (2005). "ESL Preschoolers' English Vocabulary Acquisition from Storybook Reading." *Reading Research Quarterly*, 40(4), pp. 406–408.

106 Connie Juel points out an essential linguistic issue   C. Juel (2005). "The Impact of Early School Experiences on Initial Reading." In D. Dickinson and S. Neuman, eds., *Handbook of Early Literacy Research*. New York: Guilford, Vol. 2.

106 For five years, they "learned to ignore them"   Ibid., p. 19.

106 The neuroscientist Laura-Ann Petito of Dartmouth   L-A. Petito and K. Dunbar (in press). "New Findings from Educational Neuroscience on Bilingual Brains, Scientific Brains, and the Educated Mind." In K. Fischer and T. Katzir, eds., *Building Usable Knowledge in Mind, Brain, and Education*. Cambridge University Press.

107 listening to English storybooks   Collins, "ESL Preschoolers' Vocabulary Acquisition from Storybook Reading."

CHAPTER 5: THE "NATURAL HISTORY" OF READING DEVELOPMENT

108 "No one ever told us"   Adrienne Rich (1978). "Transcendental Etude." In *The Dream of a Common Language*. New York: Norton, pp. 43–50.

108 "In a sense it is as if the child"   J. Chall (1983). *Stages of Reading Development*. New York: McGraw-Hill, p. 16.

108 Proust's extraordinary novel   M. Proust, 1981. *Remembrance of Things Past*, trans. C. K. Scott Moncrieff, Terence Kilmartin, and Andreas Mayor. New York: Random House, Vol. 1.

109 the "natural history" of the reading life   A. Bashir and A. Strominger, (1996). "Children with Developmental Language Disorders: Outcomes, Persistence, and Change." In M. Smith and J. Damico, eds., *Childhood Language Disorders*. New York: Thieme, pp. 119–140.

109 "When you learn to read"   Quoted in C. Moorehead (2000). *Iris Origo: Marchesa of Val d'Orcia*. Boston, Mass.: Godine.

109 "In books I have traveled"   A. Quindlen (1998). *How Reading Changed My Life*. New York: Ballantine, p. 6.

109 "It was my father's wish"   J. Kincaid (1996). *The Autobiography of My Mother*. New York: Farrar, Straus and Giroux. p. 12.

113 Phonological development   S. Brady (1991) "The Role of Working Memory in Reading Disability." In S. Brady and D. Shankweiler, eds., *Phonological Processes in Literacy: A Tribute to Isabelle Liberman*. Hillsdale, N.J.: Lawrence Erlbaum, pp. 129–152.

113 Morphological development   J. F. Carslisle and C. A. Stone (2005). "Exploring the Role of Morphemes in Word Reading." *Reading Research Quarterly*, 40(4), pp. 428–449. See also P. Bowers (2006). "Gaining Meaning from Print: Making Sense of English Spelling." Unpublished manuscript.

113 "as if learning natural history or music"   Rich, "Transcendental Etude," p. 43.

114 "There, perched on a cot"   J.-P. Sartre (1981). *The Words*. New York: Vintage, p. 48.

114 Each type represents dynamic changes in reading   My conceptualization owes a great deal to many theorists, and particularly to Jeanne Chall's

framework, *The Stages of Reading Development.* I am also indebted to Kurt Fischer's work on dynamic processes in reading; and to Uta Frith and Linnea Ehri and their different frameworks. Finally, in eschewing formal stages I am ultimately most similar to Perfetti. In Perfetti's nonstage framework, "many types of knowledge are acquired gradually, on the basis of many experiences." See Chall, *Stages of Reading Development.* L. C. Ehri (1998). "Grapheme-Phoneme Knowledge Is Essential for Learning to Read Words in En glish." In I. L. Metsala and L. C. Ehri, eds., *Word Recognition in Beginning Literacy.* Mahwah, N. J.: Lawrence Erlbaum, pp. 3–40. U. Frith (1985). "Beneath the Surface of Dyslexia." In K. Patterson, J. Marshall, and M. Coltheart, eds., *Surface Dyslexia.* London: Erlbaum, pp. 301–330. K. Fischer and L. T. Rose (2001). "Webs of Skill: How Students Learn." *Educational Leadership*, 59(3), pp. 6–12. K. Fischer and S. P. Rose (1998). "Growth Cycles of Brain and Mind." *Educational Leadership*, 56(3), pp. 56–60. K. Rayner, B. Foorman, C. Perfetti, D. Pesetsky, and M. Seidenberg (2001). "How Psychological Science Informs the Teacher of Reading." *Psychological Science in the Public Interest*, 2, pp. 31–74.

115 "all kinds of minds" M. Levine (1993). *All Kinds of Minds.* Cambridge, Mass.: Educators Publishing Services. M. Levine (2002). *A Mind at a Time.* New York: Simon and Schuster.

115 "Twice in your life" P. Fitzgerald (2004). "Schooldays." In T. Dooley, ed. *Afterlife.* New York: Counterpoint.

115 "I can see them standing" B. Collins (2002). "First Reader." In *Sailing Alone around the Room.* New York: Random House, p. 39.

116 "Why is it that the hardest thing" Meryl Pischa, (2001). Personal correspondence, May.

117 that each letter conveys a particular sound or two J. Downing (1979). *Reading and Reasoning.* New York: Springer Verlag.

117 large and small units of sound in the speech stream U. Goswami and P. Bryant (1990). *Phonological Skills and Learning to Read.* Hillsdale, N.J.: Lawrence Erlbaum. K. Stanovich (1986). "Matthew Effects in Reading: Some Consequences of Individual Differences in the Acquisition of Literacy." *Reading Research Quarterly*, 21(4), pp. 360–407.

117 All this, in turn, furthers reading See recent arguments and reviews of the importance and the limits of phonology's contributions to early reading in: A. Castles and M. Coltheart (2004). "Is There a Causal Link from

Phonological Awareness to Success in Learning to Read?" *Cognition, 91,* 77–111; C. Hulme, M. Snowling, M. Caravolas, and J. Carroll (2005). "Phonological Skills Are (Probably) One Cause of Success in Learning to Read: A Comment on Castles and Coltheart." *Scientific Studies of Reading,* 9(4) 351–365.

117 learn to hear and manipulate the smaller phonemes   K. Rayner et al. (2001). "How Psychological Science Informs the Teaching of Reading." J. Torgesen, R. Wagner, and C. Rashotte (1994). "Longitudinal Studies of Phonological Processing and Reading." *Journal of Learning Disabilities,* 27(10), pp. 276–286.

117 Connie Juel found that a child's phoneme awareness C. Juel (2005). "The Impact of Early School Experiences on Initial Reading." In D. Dickinson and S. Neuman, eds., *Handbook of Early Literacy Research.* New York: Guilford, Vol. 2, pp. 410–426.

118 George O. Cureton in Harlem   G. Cureton (1973). *Action-Reading.* Boston: Allyn and Bacon.

118 Children learn more easily when there are two main emphases P. E. Bryant, M. MacLean, and L. Bradley (1990). "Rhyme, Language, and Children's reading." *Applied Psycholinguistics,* 11(3), pp. 237–252.

118 A useful method for helping novice readers   Ehri, "Grapheme-Phoneme Knowledge Is Essential for Learning to Read Words in English."

118 the "sine qua non of reading acquisition"   As described extensively by the psychologist David Share of Haifa University, the self-teaching involved in reading aloud propels reading development for several reasons. Through it, young children achieve high-quality representations of words rapidly and well, thus contributing to an ever-increasing number of words in their books that are becoming part of their store of familiar words. The exact process by which children move from visual-based reading in the early phases (when they are basically memorizing the visual shape of a word like STOP) to this moment where they are forming connections between letters and pronunciations has been a matter of considerable study over the last two decades. Linnea Ehri describes this process in terms of several phases. During a pre-alphabetic phase, readers identify words by using visual cues. (Uta Frith called this period the child's logographic phase.) During Ehri's partial alphabetic phase, children learn to make "partial connections" between a word's letters and its pronunciations. Consolidating these partial connections is a major task for the novice reader, and it differs for every child. See D. Share (1995). "Phonological

Recording and Self-Teaching: Sine Qua Non of Reading Acquisition." *Cognition* 55(2), pp. 151–218. D. Share (1999). "Phonological Recording and Orthographic Learning: A Direct Test of the Self-Teaching Hypothesis." *Journal of Experimental Child Psychology,* 72(2), pp. 95–129.

118 an eminent New Zealander, the educator Marie Clay   M. Clay (1975). *What Did I Write?* Portsmouth, N.H.: Heinemann. M. Clay (1991a). *Becoming Literate: The Construction of Inner Control.* Portsmouth, N.H.: Heinemann. M. Clay (1991b). "Introducing a New Storybook to Young Readers." *Reading Teacher,* 45, pp. 264–273. M. Clay (1993). *Reading Recovery: A Guidebook for Teachers in Training.* Portsmouth, N.H.: Heinemann.

118 Irene Fountas and Gay Su Pinnell   G. Pinnell and I. Fountas (1998). *Word Matters.* Portsmouth, N.H.: Heinemann. I. C. Fountas and G. Pinnell (1996). *Guided Reading.* Portsmouth, N.H.: Heinemann.

119 Andrew Biemiller   A. Biemiller (1970). "The Development of the Use of Graphic and Contextual Information as Children Learn to Read." *Reading Research Quarterly,* 6, pp. 75–96.

120 Luckily, there are fewer irregularly spelled words   D. McGuiness (1997). *Why Our Children Can't Read and What We Can Do about It.* New York: Simon and Schuster.

121 Virginia Berninger   V. Berninger (1994). *Reading and Writing Acquisition.* Madison, Wis.: Brown and Benchmark.

121 learning about both the varied semantic meanings and common morphemes   L. C. Moats (2000). *Speech to Print: Language Essentials for Teachers.* Baltimore, Md.: Brookes.

123 As Connie Juel stresses   Juel, "The Impact of Early School Experiences on Initial Reading."

123 Vocabulary contributes to the ease and speed of decoding   G. P. Ouellette (2006). "What's Meaning Got to Do with It? The Role of Vocabulary in Word Reading and Reading Comprehension." *Journal of Educational Psychology,* 98(3), pp. 554–566.

123 As the clinician and linguist Rebecca Kennedy asserts   M. Wolf and R. Kennedy (2003). "How the Origins of Written Language Instruct Us to Teach: A Response to Steven Strauss," *Educational Researcher,* 32, pp. 26–30.

123 children have to learn about 88,700 written words   W. E. Nagy and
R. C. Anderson (1984). "How Many Words Are There in Printed School
English?" *Reading Research Quarterly*, 19(3), pp. 304–330.

124 Louisa Cook Moats calculates the sobering difference   On the first day
of school, first-grade children of higher SES levels know two to four times as
many words, on average, as children of lower SES levels. L. C. Moats (2001).
"Overcoming the Language Gap." *American Educator*, 25(5), pp. 8–9.

124 novice readers need to learn much more than the surface meaning of a
word   R. Graves and W. Slater (1987). "Development of Reading Vocabularies
in Rural Disadvantaged Students, Intercity Disadvantaged Students, and
Middle Class Suburban Students." Paper Presented at the Annual Meeting of
the American Educational Research Association, New York. E. J. Kame'enui,
R. C. Dixon, and D. W. Carnine (1987). "Issues in the Design of Vocabulary
Instruction." In M. G. McKeown, and M. E. Curtis, eds., *The Nature of
Vocabulary Acquisition*. Hillsdale, N.J.: Erlbaum, pp. 129–145.

124 Cat Stoodley's drawing   This sketch, necessarily very coarse-grained,
represents a compilation of research, particularly work on age-related
differences in the reading brain. For more specific descriptions of this work,
see the following. V. Berninger and T. L. Richards (2002). *Brain Literacy for
Educators and Psychologists*. San Diego, Calif.: Academic. J. R. Booth,
D. D. Burman, J. R. Meyer, Z. Lei, B. L. Trommer, N. D. Davenport (2003).
"Neural Development of Selective Attention and Response Inhibition."
NeuroImage, 20, pp. 737–751. E. D. Palmer, T. T. Brown, S. E. Petersen, and
B. L. Schlaggar (2004). "Investigation of the Functional Neuroanatomy of
Single Word Reading and Its Development." *Scientific Studies of Reading*,
8(3), pp. 203–223. K. R. Pugh, W. E. Mencl, A. R. Jenner, L. Katz, S. J. Frost,
J. R. Lee (2001). "Neurobiological Studies of Reading and Reading Disability."
Journal of Communication Disorders, 34, pp. 479–492. K. R. Pugh,
B. A. Shaywitz, S. E. Shaywitz, R. T. Constable, P. Skudlarski, and
R. K. Fulbright (1996). "Cerebral Organization of Component Processes in
Reading." *Brain*, 119, pp. 1221–1238. R. Sandak, W. E. Mencl, S. J. Frost,
and K. R. Pugh (2004). "The Neurobiological Basis of Skilled and Impaired
Reading: Recent Findings and New Directions." *Scientific Studies of Reading*,
8(3), pp. 273–292. B. L. Schlaggar, T. T. Brown, H. M. Lugar, K. M. Visscher,
F. M. Meizin, and S. E. Petersen (2002). "Functional Neuroanatomical
Differences between Adults and School-Age Children in the Processing of
Single Words." *Science*, 296, pp. 1476–1479. B. L. Schaggar, H. M. Lugar,
T. T. Brown, R. S. Coalson, and S. E. Petersen (2003). "fMRI Reveals Age-
Related Differences in the Development of Single Word Reading." *Society
for Neuroscience Abstracts*. B. A. Shaywitz, S. E. Shaywitz, K. R. Pugh,

W. E. Mencl, R. K. Fulbright, P. Sjudlarski (2002). "Disruption of Posterior Brain Systems for Reading in Children with Developmental Dyslexia." *Biological Psychiatry*, 52, pp. 101–110. P. Simos, J. Breier, J. Fletcher, B. Foorman, A. Mouzaki, and A. Papanicolaou (2001). "Age-Related Change in Regional Brain Activation during Phonological Decoding and Printed Word Recognition." *Developmental Neuropsychology*, 19(2), pp. 191–210. P. E. Turkeltaub, L. Gareau, D. L. Flowers, T. A. Zeffiro, and G. F. Eden (2003). "Developmental of Neural Mechanisms for Reading." *Nature Neuroscience*, 6, pp. 767–773.

125 neuroscientists at Washington University   J. A. Church, S. E. Petersen, and B. L. Schlagger (2006). "Regions Showing Developmental Effects in Reading Studies Show Length and Lexicality Effects in Adults." Poster Presented at Society for Neurosciences.

126 To be sure, adult readers activate   Palmer et al., "Investigation of the Functional Neuranatomy of Single Word Reading and Its Development."

126 "Cerebellum" means "little brain"   R. B. Ivry, T. C. Justus, and C. Middleton (2001). "The Cerebellum, Timing, and Language: Implications for the Study of Dyslexia." In M. Wolf, ed., *Dyslexia, Fluency, and the Brain.* Timonium, Md.: York, pp. 189–211. R. B. Scott, C. J. Stoodley, P. Anslow, C. Paul, J. F. Stein, E. M. Sugden, and C. D. Mitchell (2001). "Lateralized Cognitive Deficits in Children Following Cerebellar Lesions." *Developmental Medicine and Child Neurology*, 43, pp. 685–691.

127 readers need to add at least 3,000 words   Nagy and Anderson, "How Many Words Are There in Printed School English?"

128 old words become automatic, and new words come flying in   Kame'enui et al., "Issues in the Design of Vocabulary Instruction."

129 Keith Stanovich used a biblical reference   K. Stanovich (1986). "Matthew Effects in Reading: Some Consequences of Individual Differences in the Acquisition of Literacy." *Reading Research Quarterly*, 21(4), pp. 360–407.

129 has consequences for their oral and their written language   Moats, "Overcoming the Language Gap."

129 For the word-poor child, reality actually worsens   I. Beck, M. McKeown, and L. Kucan (2002). *Bringing Words to Life: Robust Vocabulary Instruction.* New York: Guildford.

129 With each step forward in reading and spelling   Carlisle and Stone, "Exploring the Role of Morphemes in Word Reading."

130 As Marcia Henry, an expert on morphology, teaches   M. Henry (2003). *Unlocking Literacy: Effective Decoding and Spelling Instruction.* Baltimore, Md.: Brookes.

130 Morphological knowledge is a wonderful dimension   V. Mann and M. Singson (2003). "Linking Morphological Knowledge to English Decoding Ability: Large Effects of Little Suffixes." In E. M. H. Assink and D. Sandra, eds., *Reading Complex Words: Cross-Language Studies.* New York: Kluwer, pp. 1–25. E. D. Reichle and C. A. Perfetti (2003). "Morphology in Word Identification: A Word Experience Model That Accounts for Morpheme Frequency Effects." *Scientific Studies of Reading,* 7, pp. 219–237.

130 "Perhaps it is only in childhood"   G. Greene (1969). *The Lost Childhood and Other Essays.* New York: Viking, p. 13.

130 I have written a great deal about fluency   M. Wolf and T. Katzir-Cohen (2001). "Reading Fluency and Its Intervention." *Scientific Studies of Reading,* 5, pp. 211–238. (Special Issue.) Specifically, my former student and now my colleague at Haifa University Tami Katzir and I suggest the following definition that will serve in its way as an outline for what needs to happen in the early phases of reading. In its beginning, reading fluency is the product of the initial development of accuracy and the subsequent development of automaticity in underlying sublexical processes, lexical processes, and their integration in single- word reading and connected text. These include perceptual, phonological, orthographic, and morphological processes at the letter, letter-pattern, and word levels, as well as semantic and syntactic processes at the word level and connected-text level. After it is fully developed, reading fluency refers to a level of accuracy and rate at which decoding is relatively effortless, oral reading is smooth and accurate with correct prosody, and attention can be allocated to comprehension.

131 The point of becoming fluent   M. Meyer and R. Felton (1999). "Repeated Reading to Enhance Fluency: Old Approaches and New Directions." *Annals of Dyslexia,* 49, pp. 83–306. R. Allington (1982). "Fluency: The Neglected Reading Goal." *Reading Teacher,* 36(6), pp. 556–561.

131 The neuroscientist Laurie Cutting of Johns Hopkins   L. Cutting and H. Scarborough (2005). "Prediction of Reading Comprehension: Relative Contribution of Word Recognition, Language Proficiency, and Other Cognitive Skills Can Depend on How Comprehension Is Measured." *Scientific Studies of Reading,* 10(3), pp. 277–299.

131 Working memory provides children with a kind of temporary space   A. Baddeley (1986). *Working Memory,* Oxford: Clarendon.

131 their comprehension becomes inextricably   Note that some of the ways we have come to learn about this relationship is through the study of children with comprehension difficulties. See, for example, the work in: K. Nation and M. Snowling (1998). "Semantic processing and the development of word recognition skills: Evidence from children with reading comprehension difficulties." *Journal of Memory and Language* 39, 85–101; J. Oakhill and N. Yuill (1996). "Higher Order Factors in Comprehension Disability: Processes and Remediation." In C. Cornaldi and J. Oakhill, eds., *Reading Comprehension Difficulties: Processes and Intervention*. Mahwah, N.J.: Erlbaum. D. Shankweiler and S. Crain (1986). "Language Mechanisms and Reading Disorder: A Modular Approach." *Cognition*, 24(1–2), 139–168. L. Swanson and J. Alexander (1997). "Cognitive Processes as Predictors of Word Recognition and Reading Comprehension in Learning-disabled and Skilled Readers: Revisiting the Specificity Hypothesis." *Journal of Educational Psychology* 89(1), pp. 128–158.

132 when children first learn to go "beyond the information given"   J. Bruner (1973). *Beyond the Information Given*. New York: Norton.

132 my Canadian colleague Maureen Lovett   M. Lovett, S. Borden, T. DeLuca, L. Lacerenza, N. Benson, and D. Brackstone. (1994). "Treating the Core Deficits of Developmental Dyslexia: Evidence of Transfer-of-Learning Following Phonologically-and-Strategy-Based Reading Training Programs." *Developmental Psychology*, 30(6), pp. 805–822. M. Lovett (2000). "Remediating the Core Deficits of Developmental Reading Disability: A Double-Deficit Perspective." *Journal of Learning Disabilities*, 33(4), pp. 334–358.

132 "At any age, the reader"   E. Bowen (1950). "Out of a Book." In *Collected Impressions*. New York: Knopf, p. 267.

CHAPTER 6: THE UNENDING STORY OF READING'S DEVELOPMENT

134 "I feel certain that"   E. Bowen (1950). "Out of a Book." In *Collected Impressions*. New York: Knopf, p. 267.

134 "I like to take my own sweet time"   T. Deeney, M. Wolf, and A. O'Rourke (1999). "I Like To Take My Own Sweet Time: Case Study of a Child with Naming-Speed Deficits and Reading Disabilities." *Journal of Special Education*, 35(3), pp. 145–155.

135 in the process he taught all of us how hard it can be   Ibid.

135 Recent reports from the National Reading Panel   P. McCardle (2001). "Emergent and Early Literacy: Current Status and Research Directions." *Learning Disabilities Research and Practice* 16(4). (Special Issue.)

135 "nation's report cards"   National Center for Education Statistics, NCES. *The Nation's Report Card: National Assessment of Educational Progress.* Washington, D.C.: U.S. Department of Education, various years. National Institute of Child Health and Human Development, NICHD. (2000). *Report of the National Reading Panel. Teaching Children to Read: An Evidence-Based Assessment of the Scientific Research Literature on Reading and Its Implications for Reading Instruction. Reports of the Subgroups.* (NIH Publication No.00-4754.) Washington, D.C.: U.S. Government Printing Office.

136 "So much of a child's life"   L. S. Schwartz (1992). "The Confessions of a Reader." In S. Gilbar, *Reading in Bed.* Jaffrey, N.H.: Godine, p. 61.

137 As the psychologist Ellen Winner describes them   E. Winner (1988). *The Point of Words: Children's Understanding of Metaphor and Irony.* Cambridge, Mass.: Harvard University Press

137 "They went off and I got aboard the raft"   M. Twain (1965). *The Adventures of Huckleberry Finn.* New York: Harper and Row, pp. 81–82.

138 The reading expert Richard Vacca   R. Vacca (2002). "From Efficient Decoders to Strategic Readers." *Reading and Writing in the Content Area,* 60(3), pp. 6–11.

139 two greatest aids to the formation of fluent comprehension   M. Pressley (2002). *Reading Instruction That Works: The Case for Balanced Teaching.* New York: Guilford.

139 For example, in reciprocal teaching   See recommendations by the following. A. S. Palincsar and A. L. Brown (1984). "Reciprocal Teaching" of Comprehension-Fostering and Comprehension-Monitoring Activities. *Cognition and Instruction* 1, pp. 117–175. A. S. Palincsar and L. R. Herrenkohl (2002). "Designing Collaborative Learning Contexts." *Theory into Practice,* 41(1), pp. 26–32. National Reading Council (2003). *Strategic Education Research Partnership.* Washington, D.C.: National Academies Press.

139 "Welcome to the *Cemetery of Dead Books*"   C. R. Zafon (2001). *The Shadow of the Wind,* trans. Lucia Graves. New York: Penguin pp. 5–6.

140 As David Rose   D. Rose (in press). "Learning in a Digital Age." In K. Fischer and T. Katzir, *Usable Knowledge.* Cambridge: Cambridge University Press.

NOTES   271

142 According to neuroscientists at Yale and Haskins Laboratory   R. Sandak et al. (2004). "The Neurobiological Basis of Skilled and Impaired Reading: Recent Findings and New Directions." *Scientific Studies of Reading*, 8(3), pp. 273–292. B. A. Shaywitz et al. (2002). "Disruption of Posterior Brain Systems for Reading in Children with Developmental Dyslexia." *Biological Psychiatry*, 52, pp. 101–110. P. E. Turkeltaub, L. Gareau, D. L. Flowers, T. A. Zettiro, and G. F. Eden (2003). "Development of Neural Mechanisms for reading." *Nature Neuroscience*, 6, pp. 767–773.

143 "And so to completely analyze"   E. B. Huey (1908). *The Psychology and Pedagogy of Reading*. Cambridge, Mass.: MIT Press, p. 6.

145 Michael Posner and various cognitive neuroscientists.   M. I. Posner and B. D. McCandliss (1999). "Brain Circuitry during Reading." In R. M. Klein and P. A. McMullen, eds., *Converging Methods for Understanding Reading and Dyslexia*. Cambridge, Mass.: MIT Press, pp. 305–337; see p. 316. M. I. Posner and A. Pavese (1998). "Anatomy of Word and Sentence Meaning." *Proceedings of the National Academy, of Sciences*, 95, pp. 899–905.

145 When expert readers look at a word   M. Posner and M. Raichle (1994). *Images of Mind*. New York: Scientific American Library. Some of the single best descriptions of the executive network, as well as the cognitive and neuroanatomical basis of reading for the layperson, may be found in V. Berninger and T. Richards (2002). *Brain Literacy for Educators and Psychologists*. San Diego, Calif.: Academic Press. Also see her earlier work: V. Berninger (1994). *Reading and Writing Acquisition*. Madison, Wisc.: Brown & Benchmark.

146 Cognitive scientists do not look at memory as one entity   A. Baddeley (1986). *Working Memory*. Oxford: Oxford University Press. P. A. Carpenter, M. A. Just, and E. D. Reichle (2000). "Working Memory and Executive Function: Evidence from Neuroimaging." *Current Opinion in Neurobiology*, 102, pp. 195–199. G. R. Lyon and N. A. Krasnegor, eds. (1996). *Attention, Memory, and Executive Function*. Baltimore, Md.: Brookes. D. L. Schacter (1993). "Understanding Implicit Memory: A Cognitive Neuroscience Approach." In A. F. Collins, eds., Collins, S. E. Gathercole, M. A. Conway, and P. E. Morris, eds. *Theories of Memory*. Hillsdale, N.J.: Lawrence Erlbaum. D. Schacter (1996). *Searching for Memory: The Brain, the Mind, and the Past*. New York: Basic Books. D. Schacter (2001). *The Seven Sins of Memory: How the Mind Forgets and Remembers*. Boston, Mass.: Houghton Mifflin.

147 psychologists call episodic memory   E. Tulving (1986). "Episodic and Semantic Memory: Where Should We Go from Here?" *Behavioral and Brain Sciences*, 9(3), pp. 573–577.

147 They also make a distinction between declarative memory   L. R. Squire
(1994). "Declarative and Nondeclarative Memory: Multiple Brain Systems
Supporting Learning and Memory." In D. L. Schacter and E. Tulving, eds.
Memory Systems. Cambridge, Mass.: MIT Press, pp. 203–231.

147 Working memory is what we use   Baddeley, *Working Memory*.

147 "A critical step in learning"   T. H. Carr (1999). "Trying to Understand
Reading and Dyslexia: Mental Chronometry, Individual Differences, Cognitive
Neuroscience, and the Impact of Instruction as Converging Sources of
Evidence." In R. M. Klein and P. A. McMullen, eds. *Converging Methods
for Understanding Reading and Dyslexia*. Cambridge, Mass.: MIT Press,
pp. 459–491.

148 the twentieth-century psychologist Donald Hebb   D. Hebb (1949). *The
Organization of Behavior*. New York: Wiley.

148 Keith Rayner, an expert in eye movements, points out   K. Rayner (1999).
"What Have We Learned about Eye Movements during Reading?" In R. Klein
and P. A. McMullen, eds., *Converging Methods for Understanding Reading
and Dyslexia*. Cambridge, Mass.: MIT Press.

148 the closeness of the connection between eye and mind   Ibid.

149 This is when our executive system influences the next eye
movements   Posner and McCandliss, "Brain Circuitry during Reading."

149 Stanislas Dehaene and Bruce McCandliss argue   B. McCandliss,
L. Cohen, and S. Dehaene (2003). "The Visual Word Form Area: Expertise for
Reading in the Fusiform Gyrus." *Trends in Cognitive Science*, 7, pp. 293–299.

149 These changes in visual specialization reach a zenith   Carr, "Trying to
Understand Reading and Dyslexia."

149 A cognitive neuroscience group in England disagrees   K. Pammer,
P. Hansen, M. I. Kringelbach, I. Holliday, G. Barnes, A. Hillebrand,
K. D. Singh, and P. I. Cornelissen (2004). "Visual Word Recognition: The
First Half Second." *Neuroimage*, 22, pp. 1819–1825.

150 An intriguing set of studies by Portuguese researchers   J. Morais et al.
(1979). "Does Awareness of Speech as a Sequence of Phones Arise
Spontaneously?" *Cognition*, 7, pp. 323–331.

151 Later brain scans of these two groups   K. M. Peterson, A. Reis, and
M. Ingvar (2001). "Cognitive Processing in Literate and Illiterate Subjects:

A Review of Some Recent Behavioral and Functional Neuroimaging Data." *Scandinavian Journal of Psychology*, 42(3), pp. 251–267.

152 The specific phonological skills used in reading   As discussed, there is more imaging research done of phonological processes than of any other process. For overviews of and different perspectives on how various factors and language systems affect this activation, see the following. Z. Breznitz (2006). *Fluency in Reading.* Mahwah, N.J.: Erlbaum. M. Coltheart, B. Curtis, P. Atkins, and M. Haller (1993). "Models of Reading Aloud: Dual Route and Parallel-Distributed Processing Approach." *Psychological Review*, 100(4), pp. 589–608. J. A. Fiez, D. A. Balota, M. E. Raichle, and S. E. Petersen (1999). "Effects of Lexicality, Frequency, and Spelling-to-Sound Consistency on the Functional Anatomy of Reading." *Neuron*, 24, pp. 205–218. C. A. Perfetti and D. J. Bolger (2004). "The Brain Might Read That Way." *Scientific Studies of Reading*, 8(4), pp. 293–304. Sandak et al., "The Neurobiological Basis of Skilled and Impaired Reading: Recent Findings and New Directions." K. R. Pugh et al. (1997). "Predicting Reading Performance from Neuroimaging Profiles: The Cerebral Basis of Phonological Effects in Printed Word Identification." *Journal of Experimental Psychology: Human Perception and Performance*, 2, pp. 1–20. L. H. Tan et al. (2005). "Reading Depends on Writing, in Chinese." Proceedings of the National Academy of Sciences, 102(24), pp. 8781–8785; R. A. Poldrack, A. D. Wagner, M. W. Prull, J. E. Desmond, G. H. Glover, and J. D. Gabrieli (1999). "Functional Specialization for Semantic and Phonological Processing in the Left Inferior Prefrontal Cortex." *NeuroImage*, 10, pp. 15–35.

152 By contrast, in more regular languages like German or Italian   H. Wimmer and U. Goswami (1994). "The Influence of Orthographic Consistency on Reading Development: Word Recognition in English and German Children." *Cognition*, 51(1), pp. 91–103.

152 Readers in the more regular Finnish, German, and Italian alphabets   E. Paulesu, J. F. Demonet, F. Fazio, F. McCrory, V. Chanoine, N. Brunswick (2001). "Dyslexia: Cultural Diversity and Biological Unity." *Science*, 291, pp. 2165–2167, March 16. E. Paulesu, E. McCrory, F. Fazio, L. Menoncello, N. Brunswick, S. F. Cappa, M. Cotelli, G. Cossu, F. Corte, M. Lorusso, S. Pesenti, A. Gallagher, D. Perani, C. Price, C. Frith, and U. Frith (2000). "A Cultural Effect on Brain Function." *Nature Neuroscience*, 3, pp. 91–96.

152 The same general principle applies to Chinese and Japanese kanji readers   Tan et al. "Reading depends on Writing, in Chinese." M. S. Kobayashi, C. W. Hayes, P. Macaruso, P. E. Hook, and J. Kato (2005). "Effects of Mora Deletion, Nonword Repetition, Rapid Naming, and Visual Search Performance on Beginning Reading in Japanese." *Annals of Dyslexia*, 55(1), pp. 105–125.

152 my colleague at Tufts Phil Holcomb   P. Holcomb (1993). "Semantic
Priming and Stimulus Degradation: Implications for the Role of the N400 in
Language Processing." *Psychophysiology*, 30, pp. 47–61. P. Holcomb (1988).
"Automatic and Attentional Processing: An Event-Related Brain Potential
Analysis of Semantic Priming." *Brain and Language*, 35, pp. 66–85. T. Ditman,
P. J. Holcomb, and G. R. Kuperberg (in press). "The Contributions of Lexico-
Semantic and Discourse Information to the Resolution of Ambiguous
Categorical Anaphors." *Language and Cognition Processes.*

153 Just as in the earlier phases in childhood   C. A. Perfetti (1985). *Reading
Ability.* New York: Oxford University Press. I. L. Beck, C. A. Perfetti, and
M. G. McKeown (1982). "Effects of Long-Term Vocabulary Instruction on
Lexical Access and Reading Comprehension." *Journal of Educational
Psychology*, 74(4), pp. 506–521.

153 Anne Fadiman   A. Fadiman (1998). *Confessions of a Common Reader.*
New York: Farrar, Straus, and Giroux.

154 Finnish researchers found that the upper temporal lobe regions
R. Salmelin and P. Helenius (2004). "Functional Neuro-Anatomy of Impaired
Reading in Dyslexia." *Scientific Studies of Reading*, 8(4), pp. 257–272.

154 the "richer" the semantic "neighborhood"   L. Locker, Jr., G. B. Simpson,
and M. Yates (2003). "Semantic Neighborhood Effects on the Recognition of
Ambiguous Words." *Memory and Cognition,* 31(4), pp. 505–515.

154 Words like "bear" and "bow"   L. Osterhout and P. Holcomb (1992).
"Event-Related Brain Potentials Elicited by Syntactic Anomalies." *Journal of
Memory and Language*, 31, pp. 285–806.

154 Syntactic information is intrinsically connected   See the discussion of
relationships between syntax and semantic processes in R. Jackendoff (2002).
Foundations of Language. Oxford: Oxford University Press.

154 Figure 6-6 shows This figure, like the time line, is a composite of
much work by many research labs, particularly the work by: J. B. Demb,
R. A. Poldrack, and J. D. Gabrieli (1999). "Functional Neuroimaging of
Word Processing in Normal and Dyslexic Readers." In R. M. Klein and
P. A. McMullen (eds.), *Converging Methods for Understanding Reading
and Dyslexia.* Cambridge, Mass.: MIT Press. P. G. Simos, J. M. Fletcher,
B. R. Foorman, D. J. Francis, E. M. Castillo, R. N. Davis, M. Fitzgerald,
P. G. Mathes, C. Denton, and A. C. Papanicolaou (in press). "Brain Activation
Profiles During the Early States of Reading Acquisition." E. D. Palmer,

T. T. Brown, S. E. Petersen, and B. L. Schlaggar (2004). "Investigation of the Functional Neuroanatomy of Single Word Reading and Its Development." *Scientific Studies of Reading* 8(3) pp. 203–223. P. Simos, J. Breier, J. Fletcher, B. Foorman, A. Mouzaki, and A. Papanicolaou (2001). "Age-Related Change in Regional Brain Activation during Phonological Decoding and Printed Word Recognition." *Developmental Neuropsychology* 19(2), pp. 191–210. K. Pammer, P. C. Hansen, M. L. Kringelbach, I. Holliday, G. Barnes, A. Hillebrand, K. D. Singh, and P. L. Cornelissen, (2004). "Visual Word Recognition: The First Half Second." *Neuroimage, 22,* pp. 1819–1825.

155 "Reading is experience"   J. Epstein (1985). "The Noblest Distraction." In *Plausible Prejudices: Essays on American Writing.* London: Norton, p. 395.

155 "For every thinking person"   H. Hesse, "The Magic of the Book." trans. D. Lindley (1974). Quoted in S. Gilbar, ed., *Reading in Bed.* Jaffrey, N.H.: Godine, p. 53.

156 "How was it that"   G. Eliot (1871, 2000). *Middlemarch.* New York: Penguin, p. 51.

157 "He had formerly observed"   Ibid.

158 "It is this demand for a *universality* of worship"   F. Dostoyevsky (1994). *The Brothers Karamazov,* trans. Ignat Avsey. Oxford: Oxford University Press, pp. 318–319.

160 research team at Carnegie Mellon   R. Mason and M. Just (2004). "How the Brain Processes Causal Inferences in Text: A Theoretical Account of Generation and Integration Component Processes Utilizing Both Cerebral Hemispheres." *Psychological Science,* 15(1), pp. 1–7. T. Keller, P. Carpenter, and M. Just (2001). "The Neural Bases of Sentence Comprehension: A fMRI Examination of Syntactic and Lexical Processes." *Cerebral Cortex,* 11(3), pp. 223–237.

161 Like Frodo, expert readers use   D. Caplan (2004). "Functional Neuroimaging Studies of Written Sentence Comprehension." *Scientific Studies of Reading,* 8(3), pp. 225–240.

161 this frontal area interacts with Wernicke's area   P. Helenius, R. Salmolin, E. Service, and J. F. Connolly (1998). "Distinct Time Course of Word and Sentence Comprehension in the Left Temporal Cortex." *Brain* 121, pp. 1133–1142.

162 "truth breaks forth, fresh and green"   A. Rich (1977). "Cartographies of Silence." In *The Dream of a Common Language.* New York: Norton, p. 20.

CHAPTER 7: DYSLEXIA'S PUZZLE AND THE BRAIN'S DESIGN

163 "For reading and writing, three years or so"   Plato. "Laws." In
E. Hamilton and E. Cairns, eds. (1961). *The Collected Dialogues*. Princeton,
N. J.: Princeton University Press, p. 810B.

165 "The greatest terror"   J. Steinbeck (1952). *East of Eden*. New York:
Putnam Penguin, pp. 270–271.

165 "I would rather clean the mold around the bathtub than read."
M. J. Adams (1990). *Beginning to Read: Thinking and Learning about Print*.
Cambridge, Mass.: MIT Press, p. 5. Quoted in C. Juel (1988). "Learning to
Read and Write: a Longitudinal Study of 54 Children from First through
Fourth Grade." *Journal of Educational Psychology*, 80, pp. 437–447.

165 "You will never understand"   J. Stewart (2001). Presentation to British
Dyslexia Associations, Sheffield, England.

167 What's missing, ironically, is a single, universally accepted definition of
dyslexia   First, consider the definition by the British Psychological Society:
"Dyslexia is evident when accurate and fluent word reading and/or spelling
develops very incompletely or with great difficulty." British Psychological
Society (1999). *Dyslexia, Literacy, and Psychological Assessment*. Leicester:
BPS, p. 18.

The International Dyslexia Association has a more specific definition:
"Dyslexia is a specific learning disability that is neurological in origin. It is
characterized by difficulties with accurate and/or fluent word recognition and
by poor spelling and decoding abilities. These difficulties typically result from
a deficit in the phonological component of language that is often unexpected
in relation to other cognitive abilities and the provision of effective classroom
instruction. Secondary consequences may include problems in reading
comprehension and reduced reading experience that can impede growth of
vocabulary and background knowledge."

The question of what constitutes dyslexia and what causes it is far from
resolved. See R. Lyon, S. Shaywitz, and B. Shaywitz (2003). "A Definition
of Dyslexia." *Annals of Dyslexia*, 52, pp. 1–14. One controversy in
defining dyslexia concerns whether the reading level of the child is or is not
commensurate with his or her IQ. Some earlier definitions specified that the
reading problems could not be the result of a poor environment, emotional,
or neurological conditions, or level of intelligence—these were referred to as
exclusionary criteria. For some time dyslexia was diagnosed only if there was
a carefully defined discrepancy between reading level and IQ, unexplainable
by these criteria.

The use of "IQ discrepancy" in the definition and diagnosis of dyslexia became the target of a long series of challenges by many eminent reading researchers. These researchers have raised many questions. For example, how accurately can an IQ test measure the verbal skills of children from linguistically impoverished environments? If a discrepancy between reading and IQ is valuable information, should it be based on the total IQ score (which includes the impact of dyslexia), or on special nonlinguistic individual IQ scores, or a discrepancy between verbal and performance scores? If the teaching methods to be used for children with dyslexia are going to be the same as those used for children with other, non-discrepancy-based problems in reading, why use discrepancy in the first place? Should children with reading problems that are not discrepancy-based receive intensive language intervention to enhance their vocabulary, along with reading intervention? If the use of "IQ discrepancy" should simply be abandoned, what would happen to the "classic discrepancy" cases? These children, who are two or more years below their real potential, usually manage (with a great deal of invisible effort) to read at grade level, and thus have no "documentable" need for services. Will abandoning discrepancy as a definition penalize children who have a "classic dyslexia" discrepancy? These questions have led to a collective effort to find better ways to define the various types of reading-impaired children. We are not there yet.

Another issue is response to intervention by children with reading disabilities. In some schools, diagnosis depends on a failure to respond to otherwise adequate intervention. See L. Fuchs and D. Fuchs (1998). "Treatment Validity: A Simplifying Concept for Reconceptualizing the Identification." *Learning Disabilities Research and Practice*, 4, pp. 204–219.

Yet another topic, emphasized in this book, concerns neurobiological origins. See B. McCandliss and K. Noble (2003). "The Development of Reading Impairment." *Mental Retardation and Developmental Disabilities*, 9, pp. 196–203.

168 the British neuropsychologist Andrew Ellis   A. Ellis (1987). "On Problems in Developing Culturally Transmitted Cognitive Modules." *Mind and Language*, 2(3), pp. 242–251.

169 A great deal of recent imaging research   See reviews in the following. M. Habib (2000). "The Neurological Basis of Developmental Dyslexia: An Overview and Working Hypothesis." *Brain*, 123, pp. 2373–2399. S. Heim and A. Keil (2004). "Large-Scale Neural Correlates of Developmental Dyslexia." *European Child and Adolescent Psychiatry*, 13, pp. 125–140. McCandliss and Noble, "The Development of Reading Impairment." The following books are useful: V. Berninger and T. Richards (2002). *Brain Literacy for Educators and Psychologists*. New York: Academic Press. S. A. Shaywitz (2003). *Overcoming*

Dyslexia. New York: Knopf. M. J. Snowling (2002). "Reading Development and Dyslexia." In U. C. Goswami, ed., *Handbook of Cognitive Development.* Oxford: Blackwell, pp. 394–411.

171 the German researcher Adolph Kussmaul   A. Kussmaul (1877). *Die Störungen der Sprache: Versuch einer Pathologie der Sprache.* Leipzig: F. C. W. Vogel.

171 the French neurologist Joseph-Jules Déjerine   J. Déjerine (1892). "Contribution à l'étude anatomo-pathologique et clinique des différentes variétés de cécité verbalè. *Mém. Soc. Biol.*, 4, p. 61. This paper was the principal topic in N. Geschwind (1962–1974). "The Anatomy of Acquired Disorders of Reading." In *Selected Papers*, Dordrecht-Holland: Reidel pp. 4–19.

172 Norman Geschwind translated Déjerine's case   N. Geschwind (1965). "Disconnexion Syndromes in Animals and Man." *Brain*, 27, pp. 237–294, 585–644.

173 The reading researcher Lucy Fildes   L. Fildes (1921). "A Psychological Inquiry into the Nature of the Condition known as Congenital Word-Blindness." *Brain*, 44, pp. 286–307.

173 In 1944, the neurologist and psychiatrist Paul Schilder   P. Schilder (1944). "Congenital Alexia and Its relation to Optic Perception." *Journal of Genetic Psychology*, 65, pp. 67–88.

173 children's inability to process phonemes within words   During the 1960s, this set of abilities was called "auditory analysis" by Jerome Rosner and Dorothea Simon, who devised the Auditory Analysis Test.

173 charted a new course for the study of reading   J. Kavanagh and I. Mattingly, eds. (1972). *Language by Ear and by Eye: The Relationship between Speech and Reading.* Cambridge, Mass.: MIT Press. See especially the following. D. Shankweiler and I. Liberman, "Misreading: A Search for Causes," pp. 293–317. M. Posner, J. Lewis, and C. Conrad. "Component Processes in Reading: A Performance Analysis," pp. 159–204. P. Gough, "One Second of Reading," pp. 331–358.

173 the psychologists Isabelle Liberman and Don Shankweiler   V. Hanson, I. Liberman, and D. Shankweiler (1983). "Linguistic Coding by Deaf Children in Relation to Beginning Reading Success." *Haskins Laboratories Status Report on Speech Research 73.*

173 Liberman and Shankweiler interpreted these and other findings
I. Y. Liberman et al. (1977). "Phonetic Segmentation and Recoding in the
Beginning Reader." In A. S. Reber and D. L. Scarborough, eds., *Toward a
Theory of Reading: The Proceedings of the CUNY Conference*. Hillsdale,
N.J.: Erlbaum. K. A. Hirsh-Pasek (1981). "Phonics without Sounds: Reading
Acquisition in the Congenitally Deaf." Unpublished doctoral dissertations,
University of Pennsylvania. R. B. Katz, D. Shankweiler, and I. Y. Liberman
(1981). "Memory for Item Order and Phonetic Recording in the Beginning
Reader." *Journal of Experimental Child Psychology*, 32, pp. 474–484.

174 The experimental psychologist Frank Vellutino   F. R. Vellutino (1979).
Dyslexia: Theory and Research. Cambridge, Mass.: MIT Press. F. R. Vellutino
(1980). "Alternative Conceptualizations of Dyslexia: Evidence in Support of a
Verbal-Deficit Hypothesis." In M. Wolf, M. K. McQuillan, and E. Radwin,
eds., *Thought and Language/Language and Reading*. Cambridge, Mass.:
Harvard Educational Review, pp. 567–587. F. Vellutino and D. Scanlon (1987).
"Phonological Coding, Phonological Awareness, and Reading Ability:
Evidence from a Longitudinal and Experimental Study." *Merrill-Palmer
Quarterly*, 33, pp. 321–363.

174 There are now hundreds of phonological studies   One example is
U. Goswami et al. (2002). "Amplitude Envelope Onsets and Developmental
Dyslexia: A New Hypothesis." *Proceedings of the National Academy of
Science*, 99, pp. 10911–10916.

175 Indeed, the most important contribution   Shaywitz, *Overcoming
Dyslexia*.

175 The researchers Joseph Torgesen and Richard Wagner   J. K. Torgesen
(1999). "Phonologically Based Reading Disabilities: Toward a Coherent
Theory of One Kind of Learning Disability." In R. J. Sternberg and L. Spear-
Swerling, eds., *Perspectives on Learning Disabilities*. New Haven, Conn.:
Westview, pp. 231–262. J. K. Torgesen, C. A. Rashotte, and A. Alexander
(2001). "Principles of Fluency Instruction in Reading: Relationships with
Established Empirical Outcomes." In M. Wolf, ed., *Dyslexia, Fluency, and the
Brain*. Timonium, Md.: York, pp. 333–355. J. K. Torgesen et al. (1999).
"Preventing Reading Failure in Young Children with Phonological Disabilities:
Group and Individual Responses to Instruction." *Journal of Educational
Psychology*, 91, pp. 579–593. J. K. Torgesen (2004). "Lessons Learned from
Research on Interventions for Students who Have Difficulty Learning to Read."
In P. McCardle and V. Chabra, eds., *The Voice of Evidence in Reading
Research*. Baltimore, Md.: Brookes, pp. 355–38.

175 Phonological research thus represents   M. W. Lovett, L. Lacerenza,
S. L. Borden, J. C. Frijters, K. A. Steinbach, and M. DePalma (2000).

"Components of Effective Remediation for Developmental Reading
Disabilities: Combining Phonologically and Strategy- Based Instruction to
Improve Outcomes." *Journal of Educational Psychology*, 92, pp. 263–283.
National Institute of Child Health and Human Developments, NICHD (2000).
*Report of the National Reading Panel. Teaching Children to Read: An
Evidence-Based Assessment of the Scientific Research Literature on Reading
and Its Implications for Reading Instruction—Reports of the Subgroups.* (NIH
Publication No.00-4754.) Washington, D.C.: U.S. Government Printing Office.
R. K. Olson, B. Wise, M. Johnson, and J. Ring (1997). "The Etiology and
Remediation of Phonologically Based Word Recognition and Spelling
Disabilities: Are Phonological Deficits the 'whole' story?" In B. Blachman,
ed., *Foundations of Reading Acquisition and Dyslexia: Implications for
Early Intervention.* Mahwah, N.J.: Lawrence Erlbaum. F. Ramus (2001).
"Outstanding Questions about Phonological Processing in Dyslexia."
Dyslexia, 7, pp. 197–216. Shaywitz, *Overcoming Dyslexia*. P. Simos, J. Breier,
J. Fletcher, B. Foorman, A. Mouzaki, and A. Papanicolaou (2001). "Age-
Related Changes in Regional Brain Activation during Phonological Decoding
and Printed Word Recognition." *Developmental Neuropsychology*, 19(2),
pp. 191–210. P. G. Simos, J. Breier, J. Fletcher, B. Foorman, E. Bergman,
K. Fishbeck, and A. Papanicolaou (2000). "Brain Activation Profiles in
Dyslexic Children during Non-Word Reading: A Magnetic Source Imagery
Study." *Neuroscience Letters*, 290, pp. 61–65. Snowling, "Reading
Development and Dyslexia." B. W. Wise, J. Ring, and R. K. Olson (1999).
"Training Phonological Awareness with and without Explicit Attention to
Articulation." *Journal of Experimental Child Psychology*, 72, pp. 271–304.

175 executive processes   H. L. Swanson (2000). "Working Memory, Short-
Term Memory, Speech Rate, Word Recognition, and Reading Comprehension
in Learning Disabled Readers: Does the Executive System Have a Role?"
Intelligence, 28, pp. 1–30. T. Gunter, S. Wagner, and A. Friederici (2003).
"Working Memory and Lexical Ambiguity Resolution as Revealed by ERPS:
A Difficult Case for Activation Theories." *Journal of Cognitive Neuroscience*,
15, pp. 43–65.

175 Virginia Berninger   V. Berninger and T. Richards (2002). *Brain Literacy
for Educators and Psychologists.* New York: Academic Press.; V. Berninger,
R. Abbott, J. Thomason, R. Wagner, H. L. Swanson, E. Wijsman, and
W. Raskind (2006). "Modeling Developmental Phonological Core Deficits
within a Working-Memory Architecture in Children and Adults with
Developmental Dyslexia." *Scientific Studies in Reading*, 10, pp. 165–198.

176 Voilà: The sum of these hypotheses   D. Bolger, C. Perfetti, and
W. Schneider (2005). "Cross-Cultural Effect on the Brain Revisited: Universal

Structures Plus Writing System Variation." *Human Brain Mapping*, 25, pp. 92–104.

176 there is not enough time for the various parts in the reading circuit For a more elaborated discussion see: D. LaBerge and J. Samuels (1974). "Toward a Theory of Automatic Information Processing in Reading." *Cognitive Psychology*, 6, pp. 293–323; C. Perfetti (1985). *Reading Ability*. New York: Oxford University Press. M. Wolf and T. Katzir- Cohen (2001). "Reading Fluency and Its Interventions." *Scientific Studies of Reading*, 5, pp. 211–238. (Special Issue).

177 Bruno Breitmeyer and the Australian researcher William Lovegrove B. G. Breitmeyer (1980). "Unmasking Visual Masking: A Look at the 'Why' Behind the Veil of 'How.' " *Psychological Review*, 87(1), pp. 52–69; W. J. Lovegrove and M. C. Williams (1993). *Visual Processes in Reading and Reading Disabilities*. Hillsdale, N.J.: Lawrence Erlbaum.

177 many children with language impairments require longer intervals P. Tallal and M. Piercy (1973). "Developmental Aphasia: Impaired Rate of Nonverbal Processing as a Function of Sensory Modality." *Neuropsychologia*, 11, pp. 389–398.

177 these difficulties in children are compounded by factors See, for example, C. Stoodley, P. Hill, J. Stein, and D. Bishop (2006). "Do Auditory Event-Related Potentials Differ in Dyslexics Even When Auditory Discrimination Is Normal?" Poster presentation at society of *Neurosciences*.

177 Usha Goswami of Cambridge U. Goswami (2003). "How to Beat Dyslexia." *Psychologist*, 16(9), pp. 462–465.

177 After observing children try to tap out rhythmic patterns P. H. Wolff (2002). "Timing Precision and Rhythm in Developmental Dyslexia." *Humanities, Social Sciences, and Law*, 15 (1–2), pp. 179–206.

177 "temporally ordered larger ensembles" P. Wolff (1993). "Impaired Temporal Resolution in Developmental Dyslexia." In P. Tallal, A. M. Galaburda, R. R. Llinas, and C. von Euler, eds. "Temporal Information Processing in the Neurons System: Special References to Dyslexia and Dysphasia." *Annals of the New York Academy of Sciences*, 682, p. 101.

178 The Israeli psychologist Zvia Breznitz Almost all of Breznitz's many studies are summarized in Z. Breznitz (2006). *Fluency in Reading*. Mahwah, N.J.: Lawrence Erlbaum.

178 what Breznitz calls an "asynchrony" This is what Charles Perfetti described as "asynchronous word processing, the failure of processing events to have been completed in time for subsequent events to use their output." In other words, if there is an asynchrony or mismatch in the time that visual information gets integrated with phonological representations, then automatic grapheme-phoneme integration, the heart of the alphabetic principle, will not be achieved. This is like the first baseman who isn't synchronized with the pitcher. One possible psychological consequence is the hypoactivation of the left angular gyrus, as seen in some studies.

178 The pediatric neurologist Martha Bridge Denckla M. B. Denckla and G. Rudel (1976). "Rapid Automatized Naming (RAN): Dyslexia Differentiated from Other Leaning Disabilities." *Neuropsychologia*, 14(4), pp. 471–479. M. B. Denckla (1972). "Color-Naming Defects in Dyslexic Boys." *Cortex*, 8, pp. 164–176. M. B. Denckla and R. Rudel (1976). "Naming of Object Drawings by Dyslexia and Other Learning-Disabled Children." *Brain and Language*, 3, pp. 1–16.

179 Denckla's discovery and her work D. Amtmann, R. D. Abbott, and V. W. Berninger (in press). "Mixture Growth Models of RAN and RAS Row by Row: Insight into the Reading System at Work across Time." *Reading and Writing, an Interdisciplinary Journal*. L. Cutting, and M. B. Denckla (2001). "The Relationship of Rapid Serial Naming and Word Reading in Normally Developing Readers: An Exploratory Model." *Reading and Writing*, 14, 673–705. M. A. Eckert, C. M. Leonard, T. L. Richard, E. H. Aylward, J. Thomas, and V. W. Berninger (2003). "Anatomical Correlates of Dyslexia: Frontal and Cerebellar Findings. *Brain*, 126 (2), pp. 482–494. K. Hempenstall (2004). "Beyond Phonemic Awareness." *Australian Journal of Learning Disabilities*, 9, pp. 3–12. C. Ho, D. W. Chan, S. Lee, S. Tsang, and V. Luan (2004). "Cognitive Profiling and Preliminary Subtyping in Chinese Developmental Dyslexia." *Cognition*, 91, 43–75. G. W. Hynd, S. R. Hooper, and T. Takahashi (1998). "Dyslexia and Language-Based Disabilities." In C. E. Coffey and R. A. Brumback, eds. *Textbook of Pediatric Neuropsychiatrists*. Washington, D.C.: American Psychiatric Press, pp. 691–718. M. Kobayashi, C. Haynes, P. Macaruso, P. Hook, and J. Kato (2005). "Effects of Mora Deletion, Nonword Repetition, Rapid Naming, and Visual Search Performance on Beginning Reading in Japanese." *Annals of Dyslexia*, 55, pp. 105–128. T. Korhonen (1995). "The Persistence of Rapid Naming Problems in Children with Reading Disabilities: A Nine-Year Follow-Up." *Journal of Learning Disabilities*, 28, pp. 232–239. H. Lyytinen (2003). Presentation of Finnish Longitudinal Study Data, International Dyslexia Association, Philadelphia, Pa., October. F. R. Manis, M. S. Seidenberg, and L. M. Doi (1999). "See Dick RAN: Rapid Naming and the Longitudinal Prediction of Reading Subskills in

First- and Second-Graders." *Scientific Studies of Reading*, 3, pp. 129–157.
C. McBride-Chang and F. Manis (1996). "Structural Invariance in the
Associations of Naming Speed, Phonological Awareness, and Verbal
Reasoning in Good and Poor Readers: A Test of the Double-Deficit
Hypothesis." *Reading and Writing*, 8, pp. 323–339. R. I. Nicolson,
A. J. Fawcett, and P. Dean (1995). "Time Estimation Deficits in Developmental
Dyslexia: Evidence of Cerebellar Involvement." *Proceedings: Biological
Sciences*, 259 (1354), pp. 43–47. R. I. Nicolson and A. J. Fawcett (1990).
"Automaticity: A New Framework for Dyslexia Research?" *Cognition*, 35 (2),
pp. 159–182. H. Swanson, G. Trainen, D. Necoechea, and D. Hammill (2003).
"Rapid Naming, Phonological Awareness, and Reading: A Meta-analysis of
the Correlation Literature." *Review of Educational Research*, 73, pp. 407–440.
L-H. Tan, J. Spinks, G. Eden, C. Perfetti, and W. T. Siok (2005) "Reading
Depends on Writing in Chinese." *PNAS*, 102, pp. 8781–8785. K. P. Van den
Bos, B. J. H. Zijlstra, and H. C. Lutje Spelberg (2002). "Life-Span Data on
Continuous-Naming Speeds, of Numbers, Letters, Colors, and Pictures
Objects, and Word-Reading Speed." *Scientific Studies of Reading*, 6,
pp. 25–49. P. F. De Jong and A. van der Leij (1999). "Specific Contributions of
Phonological Abilities to Early Reading Acquisition: Results from a Dutch
Latent-Variable Longitudinal Study." *Journal of Educational Psychology*, 91,
pp. 450–476. D. Waber (2001). "Aberrations in Timing in Children with
Impaired Reading: Cause, Effect, or Correlate?" In M. Wolf, ed. *Dyslexia,
Fluency, and the Brain*. Extraordinary Brain Series. Baltimore, Md.: York
Press, p. 103. H. Wimmer and H. Mayringer (2002). "Dysfluent Reading
in the Absence of Spelling Difficulties: A Specific Disability in Regular
Orthographies." *Journal of Educational Psychology*, 94, pp. 272–277.
M. Wolf and P. Bowers (1999). "The 'Double-Deficit Hypothesis' for the
Developmental Dyslexias." *Journal of Educational Psychology*, 91, pp. 1–24.
M. Wolf, P. G. Bowers, and K. Biddle (2000). "Naming-Speed Processes,
Timing, and Reading: A Conceptual Review." *Journal of Learning Disabilities*,
3, pp. 387–407 (Special issue).

179 "rapid automatized naming" (RAN) tasks   Years ago, my Georgia State
colleague Robin Morris, Swiss educator Heidi Bally, and I began a five-year
longitudinal exploration of the development of naming speed in children with
and without developmental dyslexia. By studying these children until the
fourth grade, and then looking back to see how children with dyslexia had
fared, we got several surprises. For the children who would later have reading
disabilities, differences in naming speed were staring us in the face from the
first days of kindergarten. These kids could not name symbols quickly. Any
symbol. But especially letters. The majority of children with severe reading
disabilities entered school with both a retrieval-speed problem (usually
undetected in oral speech), and a particular difficulty in speed of processing

with letters and the more cognitively demanding naming-speed task with set-switching (RAS). The RAS test was highly predictive of the very most impaired children in kindergarten who could not complete the set-switching RAS task, even though they could name their letters and numbers individually. We now know from many researchers that these differences in naming or retrieval speed are found throughout childhood and continue into adulthood. We know that general naming ability in children as young as three years of age can predict some forms of later reading disability, and other learning disabilities such as attention deficit disorder. See, for example, Rosemary Tannock's excellent work on the intriguing differences in color and object naming that children with only attention issues have: R. Tannock, R. Martinussen, and J. Frijters (2000). "Naming Speed Performance and Stimulant Effects Indicate Effortful, Semantic Processing, Deficits in Attention Deficit/Hyperactivity Disorder." *Journal of the American Academy of child and Adolescent Psychiatry*, 28, pp. 237–252. See also M. Wolf (1986). "Rapid Alternating Stimulus (R.A.S.) Naming: A Longitudinal Study in Average and Impaired Readers." *Brain and Language*, 27, pp. 360–379; M. Wolf and M. Denckla (2005) RAN/RAS Tests, Pro-Ed Publishers. RANRAS: Rapid Automatized Naming and Rapid Alternating Stimulus Test. Austin, Tex.: Pro-Ed.

179 Just as Geschwind suspected   N. Geschwind (1965). "Disconnexion Syndromes in Animals and Man."

179 brain images of naming speed   M. Misra, T. Katzir, M. Wolf, R. and A. Poldrack (2004). "Neural Systems for Rapid Automatized Naming in Skilled Readers: Unraveling the RAN-Reading Relationship." *Scientific Studies in Reading*, 8(3), pp. 241–256.

179 Just as was hypothesized   B. McCandliss, L. Cohen, and S. Dehaene (2003). "Visual word form area: Expertise for reading in the fusiform gyrus." *Trends in Cognitive Science*, 7, pp. 293–299.

181 This is why measures of naming speed like RAN and RAS predict reading   G. DiFilippo, D. Brizzolara, A. Chilosi, M. DeLuca, A. Judica, C. Pecini, D. Spinell, and P. Zoccolotti (in press). "Naming Speed and Visual Search Deficits in Disabled Readers: Evidence from an Orthographically Regular Language"; see also V. Närhi, T. Ahonen, M. Aro, T. Leppäsaari, T. Korhonen, A. Tolvanen, and H. Lyytinen (2005). "Rapid Serial Naming: Relations between Different Stimuli and Neuropsychological Factors." *Brain and Language*, 92, pp. 45–57.

181 We know that the great majority of children with dyslexia
P. T. Ackerman, R. A. Dykman, and M. Y. Gardner (1990). "Counting Rate,

Naming Speed, Phonological Sensitivity, and Memory Span: Major Factors in Dyslexia." *Journal of Learning Disabilities*, 23, pp. 325–337. D. Amtmann, R. Abbott, and V. Berninger (in press). "Mixture Growth Models of RAN and RAS Row by Row: Insight into the Reading System at Work across Time." *Reading and Writing*. N. Badian (1995). "Predicting Reading Ability over the Long Term: The Changing Roles of Letter Naming, Phonological Awareness, and Orthographic Knowledge." *Annals of Dyslexia*, 45, pp. 79–86. D. Compton (2000). "Modeling the Relationship between Growth in Rapid Naming Speed and Growth in Decoding Skill in First Grade Children." *Journal of Educational Psychology*, 95, pp. 225–239. DiFilippo et al. "Naming Speed and Visual Search Deficits in Disabled Readers." U. Goswami et al. (2002). "Amplitude Envelope Onsets and Developmental Dyslexia: A New Hypothesis." *PNAS*, 99, pp. 10911–10916. J. Kirby, R. Parilla, and S. Pfeiffer (2003). "Naming Speed and Phonological Awareness as Predictors of Reading Development." *Journal of Educational Psychology*, 95(3), pp. 453–464. K. Pammer, P. Hanson, M. Kringlebach, I. Holliday, G. Barnes, A. Hillebrand, K. Singh, and P. Cornelissen (2004). "Visual Word Recognition: The First Half Second." *Neuroimaging*, 22, pp. 1819–1825. H. Swanson, G. Trainen, D. Necoechea, and D. Hammill (2003). "Rapid Naming, Phonological Awareness, and Reading: A Meta-Analysis of the Correlation Literature." *Review of Educational Research*, 73, pp. 407–440. M. Wolf, H. Bally, and R. Morris (1986). "Automaticity, Retrieval Processes, and Reading: A Longitudinal Study in Average and Impaired Readers." *Child Development*, 57, pp. 988–1000.

182 resurrected the concept of the "disconnexion syndrome"   Geschwind, "Disconnexion Syndromes in Animals and Man."

182 The two most common ideas located the source of failure   M. Blank and W. H. Bridger (1964). "Cross-Modal Transfer in Nursery School Children." *Journal of Comparative and Physiological Psychology*, 58, pp. 277–282. H. Birch and L. Belmont (1964). "Auditory-Visual Integration in Normal and Retarded Readers." *American Journal of Orthopsychiatry*, 34, pp. 852–861.

182 Modern neuroscience goes below the surface   See, for example, K. Pugh et al. (2000). "The Angular Gyrus in Developmental Dyslexia: Task Specific Differences in Functional Connectivity in Posterior Cortex." *Psychological Science*, 11, pp. 51–59.

182 Italian neuroscientists   E. Paulesu, U. Frith, M. Snowling, A. Gallagher, J. Morton, and R. S. J. Frackowiak (1996). "Is Developmental Dyslexia a Disconnection Syndrome? Evidence from PET Scanning." *Brain*, 119, pp. 143–157. E. Paulesu, J. Demonet, F. Fazio, E. McCrory, V. Chanoine, N. Brunswick,

S. Cappa, G. Cossu, M. Habib, C. Frith, and U. Frith (2001). "Dyslexia: Cultural Diversity and Biological Unity." *Science*, 291, pp. 2165–2167.

182 in an expansive connecting area called the "insula"   E. Paulesu et al., "Is Developmental Dyslexia a Disconnection Syndrome?"

182 Researchers from Yale   S. Shaywitz, B. Shaywitz, W. E. Mencl, R. K. Fulbright, P. Skudlarski, R. T. Constable, K. Pugh, J. Holahan, K. Marchione, J. Fletcher, G. R. Lyone, and J. Gore (2003). "Disruption of Posterior Brain Systems for Reading in Children with Developmental Dyslexia." *Biological Psychiatry*, 52, pp. 101–110.

183 they found that this area 37 is not connected   See discussion of functional connectivity in R. Sandak, W. E. Mencl, S. J. Frost, and K. R. Pugh (2004). "The Neurological Basis of Skilled and Impaired Reading: Recent Findings and New Directions." *Scientific Studies of Reading*, 8(3), pp. 273–292.

183 some neuroscientists find   B. Horwitz, J. Rumsey, and B. Donohue (1998). "Functional Connectivity of the Angular Gyrus in Normal Reading and Dyslexia." *Proceedings of the National Academy of Sciences*, 95, pp. 8939–8944.

183 A research group in Houston   P. G. Simos, J. Breier, J. Fletcher, B. Foorman, E. Bergman, K. Fishbeck, and A. Papanicolaou (2000). "Brain Activation Profiles in Dyslexic Children during Non-Word Reading. A Magnetic Source Imagery Study." *Neuroscience Letters*, 290, pp. 61–65.

183 some of my colleagues at MIT   J. D. E. Gabrieli, R. A. Poldrack, and J. E. Desmond (1998). "The Role of Left Prefrontal Cortex in Language and Memory." *Proceedings of National Academy of Sciences*, 95(3), pp. 906–913.

183 Samuel T. Orton and his colleague Anna Gillingham   S. Orton (1928). "Specific Reading Disability—Strephosymbolia." *Journal of the American Medical Association*, 90, pp. 1095–1099.

184 Researchers in the 1960s and 1970s were fascinated   M. P. Bryden (1970). "Laterality Effects in Dichotic Listening: Relations with Handedness and Reading Ability in Children." *Neuropsychologia*, 8, pp. 443–450.

184 Not only was the speed of the impaired readers significantly worse   E. B. Zurif and G. Carson (1970). "Dyslexia in Relation to Cerebral Dominance and Temporal Analysis." *Neuropsychologia*, 8, pp. 351–361.

185 Similarly, in the 1970s researchers   K. Rayner and F. Pirozzolo (1977). "Hemisphere Specialization in Reading and Word Recognition." *Brain and*

Language, 4(2), pp. 248–261. K. Rayner and F. Pirozzolo (1979). "Cerebral
Organization and Reading Disability." *Neuropsychologia*, 17(5), pp. 485–491.

185 One lateralization study after another during this period   G. Yeni-
Komshian, D. Isenberg, and H. Goldberg (1975). "Cerebral Dominance and
Reading Disability: Lateral Visual Field Deficit in Poor Readers."
Neuropsychologia, 13, pp. 83–94.

185 the research group at Georgetown University found   P. Turkeltaub,
L. Gareau, L. Flowers, T. Zeffiro and G. Eden (2003). "Development of
Neural Mechanisms for Reading." *Nature Neuroscience*, 6, pp. 767–773.

185 S. Shaywitz, B. Shaywitz, K. Pugh, W. Mencl, et al. (1998). "Functional
disruption in the organization of the brain for reading in dyslexia."
Proceedings of the National Academy of Sciences, USA, 95, pp. 2636–2641.
S. Shaywitz (2003). *Overcoming Dyslexia*.

185 Researchers at Yale   S. Shaywitz, B. Shaywitz, W. E. Mencl, R. K. Fulbright,
P. Skudlarski, R. T. Constable, K. Pugh, J. Holahan, K. Marchione, J. Fletcher,
G. R. Lyon, and J. Gore. (2003). "Disruption of Posterior Brain Systems for
Reading in Children with Developmental Dyslexia." *Biological Psychiatry*, 52,
pp. 101–110.

186 This time line is a product of cumulative research   J. B. Demb,
R. A. Poldrack, and J. D. E. Gabrieli (1999). "Functional Neuroimaging of
Word Processing in Normal and Dyslexic Readers." In R. M. Klein and
P. A. McMullen eds., *Converging Methods for Understanding Reading and
Dyslexia*. Cambridge, Mass.: MIT Press. Habib, "The Neurological Basis
of Developmental Dyslexia." P. H. T. Leppanen and H. Lyytinen (1997).
"Auditory Event-Related Potentials in the Study of Developmental Language-
Related Disorders." *Auditory and Neuro-Otology*, 2, pp. 308–340. H. Lyytinen
(2003). Presentation of Finnish Longitudinal Study Data, International
Dyslexia Association: Philadelphia, Pa., October. Pammer et al., "Visual Word
Recognition: The First Half Second." J. M. Rumsey (1997). "Orthographic
Components of Word Recognition: A PET-rCBF Study." *Brain*, 120, pp. 739–
759. R. Salmelin and P. Helenius (2004). "Functional Neuro-Anatomy of
Impaired Reading in Dyslexia." *Scientific Studies of Reading*, 8(4), pp. 257–
272. Sandak et al., "The Neurobiological Basis of Skilled and Impaired
Reading: Recent Findings and New Directions." Simos et al., "Age-Related
Changes in Regional Brain Activation during Phonological Decoding and
Printed Word Recognition." Turkeltaub et al., "Developmental of Neural
Mechanisms for Reading."

188 As eminent researchers   O. Tzeng and W. S-Y. Wang (1982). "Search
for a Common Neurocognitive Mechanism for Language and Movements."

American Journal of Physiology, 246, pp. 904–911. O. Tzeng and W. S-Y. Wang (1983). "The First Two R's." *American Scientist*, 71, pp. 238–243.

188 The evocative picture of a right-hemisphere-dominated circuit   Guinevere Eden and her research team summarized some of the possible nonexclusive hypotheses about what underlies phonological difficulties in dyslexia: a disconnection in the left-hemisphere circuits between more frontal and posterior regions; problems in left frontal regions; developmental differences and weaknesses in the left temporal-parietal regions, particularly areas around the angular gyrus; and right-hemispheric reorganization to compensate for left-hemisphere weaknesses. To be sure, posterior weaknesses could cause children with dyslexia to overrely on bilateral frontal areas to compensate for what their left posterior structures aren't able to do easily or quickly. And posterior weaknesses in the left hemisphere could also help explain why right-hemisphere regions initially become more involved. In typical reading, visual information is directed to occipital areas in both hemispheres, and then the information in the right visual areas is sent across the corpus callosum to the left visual regions to be integrated with the left-lateralized orthographic and language operations. In dyslexia the left posterior weaknesses could rearrange this direction of input. As the Shaywitz group underscore, such posterior underactivation could result in less efficient, memory-intensive strategies in reading.

188 An understanding of the principles of brain design in reading   Pammer et al., "Visual Word Recognition: The First Second Half."

188 Accepting the idea of subtypes   D. Doehring, I. M. Hoshko, and M. Bryans (1979). "Statistical Classification of Children with Reading Problems." *Journal of Clinical Neuropsychology*, 1, pp. 5–16. R. Morris (1982). "The Developmental Classification of Leaning Disabled Children Using Cluster Analysis." Dissertation, University of Florida.

188 My Canadian colleague Pat Bowers and I   M. Wolf and P. Bowers (2000). "The Question of Naming-Speed Deficits in Developmental Reading Disability: An Introduction to the Double-Deficit Hypothesis." *Journal of Learning Disabilities*, 33, pp. 322–324. (Special Issue.) Wolf and Bowers, "The 'Double-Deficit Hypothesis' for the Developmental Dyslexias." P. G. Bowers and M. Wolf (1993). "Theoretical Links among Naming Speed, Precise Timing Mechanisms, and Orthographic Skill in Dyslexia." *Reading and Writing*, 5, pp. 69–85.

189 Very important, just under 20 percent of the poor readers   M. Wolf and P. G. Bowers (1999). "The Double-Deficit Hypothesis for the Developmental Dyslexias." *Journal of Educational Psychology*, 91, pp. 415–438.

189 M. W. Lovett, K. A. Steinbach, and J. C. Frijters (2000). "Remediating the Core Deficits of Developmental Reading Disability: A Double-Deficit Perspective." *Journal of Learning Disabilities*, 33(4), pp. 334–358.

189 German and Spanish languages (See work in these languages): H. Wimmer, H. Mayringer, and K. Landerl (2000); "The Double-Deficit Hypothesis and Difficulties in Learning to Read Regular Orthography." *Journal of Educational Psychology*, 92, pp. 668–680. C. Escribano (in press). "The Double-Deficit Hypothesis: Comparing the Subtypes of Children in a Regular Orthography."

189 psychologist Bruce Pennington   See the work of the cognitive and genetic psychologist Bruce Pennington at the University of Denver, who comes closest to the developmental, multiple-process view of reading development and failure described here. In his "multiple cognitive deficit" view, there can be several possible sources and manifestations of reading failure, and depending on time and intervention, these can look different over time. B. F. Pennington (2006). "From Single to Multiple Deficit Models of Developmental Disorders." *Cognition*, 101(2), pp. 385–413.

189 Robin Morris's group   R. Morris, K. Stuebing, J. Fletcher, S. Shaywitz, G. R. Lyon, D. Shankweiler et al. (1998). "Subtypes of Reading Disability: Variability around a Phonological Core." *Journal of Educational Psychology*, 90, pp. 347–373.

190 The sociolinguist Chip Gidney at Tufts   Theresa Deeney, Calvin Gidney, Maryanne Wolf, and Robin Morris (1998). "Phonological Skills of African-American Reading-Disabled Children." Paper presented at Society for the Scientific Studies of Reading.

190 the Austrian psychologist Heinz Wimmer   K. Landerl, H. Wimmer, and U. Frith (1997). "The Impact of Orthographic Consistency on Dyslexia: A German-English Comparison." *Cognition*, 63(3), pp. 315–334.

191 somewhat different uses of the major structures   K. Pugh, R. Sandak, S. Frost, D. Moore, and E. Mencl. (2005). "Examining Reading Development and Reading Disability in English Language Learners: Potential Contributions from Functional Neuroimaging." *Learning Disabilities Research and Practice*, 20, pp. 24–30. L. H. Tan, J. Spinks, G. Eden, D. Perfetti, and W. Sick (2005). "Reading Depends on Writing in Chinese." *Proceedings of National Academy of Sciences*, 102, pp. 8781–8785.

191 Researchers in Hong Kong   C. Ho, D. W. Ghen, S. Lee, S. Taang, and Luan (2004). "Cognitive Profiling and Preliminary Subtyping in Chinese Developmental Dyslexia." *Cognition*, 91, pp. 43–75.

191 Among Spanish-speakers   Escribano, "The Double-Deficit Hypothesis."

191 Similar data emerged for Hebrew   T. Katzir, S. Shaul, Z. Breznitz, and
M. Wolf (2004). "Universal and Unique Characteristics of Dyslexia: A Cross-
Linguistic Comparison of English- and Hebrew-Speaking Children."
(Unpublished research.)

191 In these more transparent languages   Escribano, "The Double-Deficit
Hypothesis." Katzir et al. "Universal and Unique characteristics of Dyslexia."
Landerl et al., "The Impact of Orthographic Consistency on dyslexia."
Paulesu et al., "Dyslexia: Cultural Diversity and Biological unity."

192 "It is said that Mr. Beckmann could neither read nor write."   Albert
Kleber, OSB, STD (1940). *Ferdinand, Indiana, 1840–1940: A Bit of Cultural
History*. Saint Meinrad, Ind., p. 67.

CHAPTER 8: GENES, GIFTS, AND DYSLEXIA

198 "If only we knew"   D. Whyte (1990) "The Faces at Braga." In *Where
Many Rivers Meet*. Langley, Wash.: Many Rivers.

199 The neuropsychologist P. G. Aaron   P. G. Aaron and R. G. Clouse (1982).
"Freud's Psychohistory of Leonardo da Vinci: A Matter of Being Right or
Left." *Journal of Interdisciplinary History*, 13(1), pp. 1–16.

199 "My principal weakness was a bad memory"   A. Einstein (1954). Letter
to Sybille Bintoff, May 21. Cited in A. Folsing (1997). *Albert Einstein*. New
York: Penguin.

199 words did "not seem to play any role"   S. F. Witelson, D. L. Kigar, and
T. Harvey (1999). "The Exceptional Brain of Albert Einstein." *Lancet*, 353,
pp. 2149–2153.

199 Whether Einstein might have met the criteria   Ibid.

199 autopsy of Einstein's brain   A dissenting view is found in A. Galaburda
(1999). "Albert Einstein's Brain." *Lancet*, 354, p. 1821.

199 discovered unexpected symmetries between the hemispheres   Witelson et
al., "The Exceptional Brain of Albert Einstein."

201 The neurologist Al Galaburda   A. M. Galaburda (2005). Personal
correspondence, November 27.

201 Over eighty years ago, Samuel Orton   S. Orton (1928). "Specific Reading Disability—Strephosymbolia." *Journal of the American Medical Association*, 90, pp. 1095–1099.

201 More than fifty years later, Norman Geschwind   N. Geschwind (1982). "Why Orton Was Right." *Annals of Dyslexia*, 32, pp. 13–28.

202 "Dyslexics themselves are frequently endowed"   Ibid., pp. 21–22.

203 In most people the planum temporale   A. Galaburda (1993). "Neuroanatomical Basis of Developmental Dyslexia." *Neurological Clinical*, 11, pp. 161–173. A. Galaburda, J. Cosiglia, G. Rosen, and G. Sherman (1987). "Planum Temporale Asymmetry: Reappraisal since Geschwind and Levitsky." *Neuropsychologia*, 25, 853–868.

203 The potential importance of this explanation lost ground   George Hynd, Lynn Flowers, and their group replicated the result of a larger RH planum in a group of individuals with dyslexia, but Stanford researchers John Gabrieli and his group then did not. The latter group speculated that this RH different might be present only in a subgroup of dyslexia, a leitmotiv in much dyslexia research. Researcher Pauline Filipek reviewed a range of asymmetry studies and concluded that the supportive evidence was too little in part because of mapping discrepancies between studies (i.e., differences about where one region begins and the other ends), a conclusion also reached by the Stanford group. P. A. Filipek (1995). "Neurobiologic Correlates of Developmental Dyslexia: How do Dyslexics' Brains Differ from Those of Normal Readers?" *Journal of Child Neurology*, 10(1), pp. 62–69. Galaburda, "Neuroanatomical Basis of Developmental Dyslexia." G. W. Hynd, M. Semrud-Clikeman, A. R. Lerys, E. S. Novey, and D. Eliopulos (1990). "Brain Morphology in Developmental Dyslexia and Attention Deficit Disorder/Hyperactivity." *Archives of Neurology*, 47, pp. 919–926.

204 the investigations at the cellular level—cells that are responsible for fast or transient processing   A. Galaburda (2006). "Dyslexia: Advances in Cross-Level Research." In G. Rosen, ed., *The Dyslexic Brain*. Mahwah, N.J.: Erlbaum. A. R. Jenner, G. D. Rosen, and A. M. Galaburda (1999). "Neuronal Asymmetries in Primary Visual Cortex of Dyslexic and Nondyslexic Brains." *Annals of Neurology*, 46, pp. 189–196.

204 Galaburda argued that cellular differences   Jenner et al., "Neuronal Asymmetries in Primary Visual Cortex of Dyslexic and Nondyslexic Brains." See also J. C. Greatrex and N. Drasdo (1995). "The Magnocellular Deficit Hypothesis in Dyslexia: A Review of Reported Evidence." *Opthalmic and Physiological Optics*, 15(5), pp. 501–506.

204 the neuroscientist Glenn Rosen   G. Rosen, ed. (2005). *The Dyslexic Brain: New Pathways in Neuroscience Discovery.* Mahwah, N.J.: Lawrence Erlbaum. G. D. Rosen et al. (2001). "Animal Models of Developmental Dyslexia: Is There a Link between Neocortical Malformations and Defects in Fast Auditory Processing?" In M. Wolf, ed., *Dyslexia, Fluency, and the Brain.* Timonium, Md.: York, pp. 129–157.

204 as a result of the lesions, the mice could no longer process   By analogy, humans who have genetically based anomalies similar to those in Glenn's mice would have difficulties whenever they were required to process rapidly presented acoustic and phonemic-level information, as in speech. If their anomalies involved visual areas, they would be impaired in processing rapidly presented visual information, as in print.

204 A study by neurologists in Boston   B. Chang, T. Katzir, C. Walsh, et al. (in press). "A Structural Basis for Reading Fluency: Cortico-Cortical Fiber Tract Disruptions Are Associated with Reading Impairment in a Neuronal Migration Disorder."

205 the many sources of problems with fluency in children with reading disabilities   M. Wolf and T. Katzir-Cohen (2001). "Reading Fluency and Its Intervention." *Scientific Studies of Reading,* 5, pp. 211–238. (Special Issue.)

206 As an eminent geneticist has observed   S. Petrill (2005). "Introduction to This Special Issue: Genes, Environment, and the Development of Reading Skills." *Scientific Studies of Reading,* 9, pp. 189–196.

206 The geneticist Elena Grigorenko of Yale   E. Grigorenko (2005). "A Conservative Meta-Analysis of Linkage and Linkage-Association Studies of Developmental Dyslexia." *Scientific Studies of Reading,* 9(3), pp. 285–316.

206 as observed by Bruce Pennington   B. F. Pennington (2006). "From Single to Multiple Deficit Models of Developmental Disorders." *Cognition,* 101(2), pp. 385–413.

206 Finnish and Swedish researchers   K. Hannula-Jouppi, N. Kaminen-Ahola, M. Taipale, P. Eklund, J. Nopola- Hommi, H. Kaariainen, and J. Kere (2005). "The Axon Guidance Receptor Gene ROBO1 Is a Candidate Gene for Developmental Dyslexia." *PLOS Genetics,* 1(4), pp. 467–474.

206 For English-speakers, researchers at Yale   H. Meng, S. D. Smith, K. Hager, M. Held, L. Liu, R. K. Olson, B. F. Pennington, J. C. DeFries, Gelernter, T. O'Reilly-Pol, S. Semlo, Skudlarski, S. E. Shaywitz, D. A. Shaywitz,

K. Marchiene, Y. Wang, M. Paramasivam, J. J. LeTuree, G. P. Page, and Gruen (2005). "*DCDC2* Is Associated with Reading Disability and Modulates Neuronal Development in the Brain." *Proceedings of National Academy of Sciences*, 102(47), pp. 17053–17058.

207 a large Finnish family with a long genetic history of dyslexia
K. Hannula-Jouppi, N. Kaminon-Ahola, M. Taipale, R. Eklund, J. Nopola-Hemmi, H. Kaariainen, and J. Kere (2005). "The Axon Guidance Receptor Gene ROBO1 Is a Candidate Gene for Developmental Dyslexia." J. Nopola-Hemmi, B. Myllyluema, A. Voutilainen, S. Leinonen, and J. Kere (2002). "Familial Dyslexia: Neurocognitive and Genetic Correlation in a Large Finnish Family." *Developmental and Medical Child Neurology*, 44, pp. 580–586.

207 Other support comes from one of the largest and most established genetic programs   R. K. Olson (2004). "SSSR, Environment, and Genes." *Scientific Studies of Reading*, 8(2), pp. 111–124. B. Byrne, C. Delaland, R. Fielding-Barnsley, P. Quain, S. Sumelsson, and T. Hoien (2002). "Longitudinal Twin Study of Early Reading Developmental in Three Countries: Preliminary Results." *Annals of Dyslexia*, 52, pp. 49–74.

209 "I am somehow less interested"   Stephen Jay Gould (1980). *The Panda's Thumb: More Reflections in Natural History*. New York: Norton.

209 And we need educational research   R. Lyon (2001). "Measuring Success: Using Assessment and Accountability to Raise Student Achievement." Statement to the Subcommittee on Education Reform, U.S. House of Representatives.

210 intervention program (RAVE-O)   M. Wolf, L. Miller, and K. Donnelly (2000). "RAVE-O: A Comprehensive Fluency-Based Reading Intervention Program." *Journal of Learning Disabilities*, 33, pp. 375–386 (special issue). R. Morris, M. Lovett, M. Wolf (submitted 2006). "The Case for Multiple-Component Remediation of Reading Disabilities: A Controlled Factorial Evaluation of the Influence of IQ, Socioeconomic Status, and Race on Outcomes" (submitted for publication).

CHAPTER 9: CONCLUSIONS
212 "Each torpid turn of the world"   R. M. Rilke (1939). "The Seventh Elegy." In *Duino Elegies*. New York: Norton, p. 63.

212 "Reading is an act."   J. Carroll (2001). "America's Bookstores: Shrines to the Truth." *Boston Globe*, p. A11, January 30.

212 "In the clash between the conventions"　K. Kelly (2006). "Scan This Book!" *New York Times Magazine*, Section 6, p. 43, May 14.

213 "We can have confidence"　R. Kurzweil (2006). *The Singular Is Near.* New York: Penguin, pp. 197–198; "How can we," p. 487.

216 A machine that can read　Ibid., p. 589. Kurzweil 3000 Reading System. Kurzweil Educatitonal Systems.

217 Charles Perfetti, Li-Hai Tan, and their group　L-H Tan et al. (2000). "Brain Activation in the Processing of Chinese Characters and Words: A Functional MRI Study." *Human Brain Mapping*, 10(1), pp. 16–27. L-H Tan et al. (2003). "Neural Systems of Second Language Reading Are Shaped by Native Language." *Human Brain Mapping*, 18(3), pp. 158–166.

217 the classicist Eric Havelock.　E. Havelock (1976). *Origins of Western Literacy.* Ontario, Canada: Ontario Institute for Studies in Education.

217 cognitive neuroscientists　M. I. Posner and B. D. McCandliss (1999). "Brain Circuitry during Reading." In R. Klein and P. McMullen, eds., *Converging Methods for Understanding Reading and Dyslexia.* Cambridge, Mass.: MIT Press.

219 Few scholars are more eloquent　W. Ong (1982). *Orality and Literacy.* London: Methuen, p. 178.

220 Only these conditions assured Socrates　Plato, "Phaedrus." In E. Hamilton and H. Cairns, eds. (1961). *The Collected Dialogues.* Princeton, N.J.: Princeton University Press, p. 276.

220 The contemporary scholar John McEneaney　J. McEneaney (2006). "Agent-Based Literacy Theory." *Reading Research Quarterly*, 41, pp. 352–371.

220 "choice in appearance"　D. Rose (in press). "Learning in a Digital Age." In K. Fischer and T. Katzir, eds., *Usable Knowledge.* Cambridge: Cambridge University Press.

221 At the start of this book I quoted the technology expert Edward Tenner　E. Tenner (2006). "Searching for Dummies." *New York Times*, Section 4, p. 12, March 26.

223 The more young children are read to　G. J. Whitehurst and C. J. Lonigan (1998). "Child Development and Emergent Literacy." *Child Development*,

69(3), pp. 848–872. G. J. Whitehurst et al. (1994). "A Picture Book Reading Intervention in Day Care and Home for Children from Low-Income Families." *Developmental Psychology*, 30, pp. 679–689. G. J. Whitehurst and C. J. Lonigan (2001). "Emergent Literacy: Development from Prereaders to Readers." In S. B. Neuman and D. K. Dickinson, eds., *Handbook of Early Literacy Research*. New York: Guilford, pp. 11–29.

224 "The rich get richer and the poor poorer." K. Stanovich (1986). "Matthew Effects in Reading: Some consequences of Individual Differences in the Acquisition of Literacy." *Reading Research Quarterly*, 21(4), pp. 360–407.

224 the fluent reading brain activates M. A. Just, P. A., Carpenter, T. A., Keller, W. F., Eddy, and K. R. Thulborn (1996). "Brain Activation Modulated by Sentence Comprehension." *Science*, 274(5284), pp. 912–913.

225 Recently I read an essay David S. Kahn (2006). "How Low Can They Go?" *Wall Street Journal*, p. W11, May 26.

227 As Norman Geschwind often asserted N. Geschwind (1982). "Why Orton Was Right." *Annals of Dyslexia*, 32, pp. 13–28.

227 hopeful applications of neuroscience Interventions based on this knowledge, such as our RAVE-O program, Lovett's PHAST program, Rose's Thinking Reader, and Breznitz's acceleration programs, are a modest but encouraging start toward what will become interventions in the future. See descriptions of RAVE-O and PHAST in the following. M. Wolf, L. Miller, and K. Donnelly (2000). "RAVE-O: A Comprehensive Fluency-Based Reading Intervention Program." *Journal of Learning Disabilities*, 33, pp. 375–386 (special issue). R. Morris et al. (submitted). "The Case for Multiple-Component Remediation of Reading Disabilities: A Controlled Factorial Evaluation of the Influence of IQ, Socioeconomic Status, and Race on Outcomes." Rose, "Learning in a Digital Age." Z. Breznitz (1997). "The Effect of Accelerated Reading Rate on Memory for Text among Dyslexic Readers." *Journal of Educational Psychology*, 89, pp. 287–299.

228 foster their resilience G. Noam and C. Herman (2002). "Where Education and Mental Health Meet: Developmental Prevention and Early Intervention in Schools." *Development and Psychopathology*, 14, pp. 861–875. C. Recklitis and G. Noam (1999). "Clinical and Developmental Perspectives on Adolescent Coping." *Child Psychiatry and Human Development*, 30, pp. 87–101.

INDEX